Advances in Plant Breeding - From Techniques to Stress Tolerance

Edited by Murat Aycan

Published in London, United Kingdom

Advances in Plant Breeding - From Techniques to Stress Tolerance
http://dx.doi.org/10.5772/intechopen.1008097
Edited by Murat Aycan

Contributors
Aditi R. Rajane, Anbuselvan Antony, Ankita Manna, Bekir Bülent Arpacı, Bhagyashree P. Bahatkar, Burak Özgören, Chinnappan Deepa Joan of Arc, Debnirmalya Gangopadhyay, Doğancan Kahya, Hajime Yanai, Hatice Nilufer Yıldız, Kiran R. Kharat, Krishnagowdu Saravanan, Madhushri Das Datta, Murat Aycan, Mustafa Yildiz, Muthukumar Elankumaran, Pakize Gök Güler, Pallavi R. Sasane, Pinar Arıdıcı Kara, Prosenjeet Chakraborthy, Rahul Chatterjee, Raveendran Pottathil, Refik Bozbuga, Ritwik Acharya, Shivaanie Nandakumar, Shubhajit Shaw, Sundarasamy Dhanapal, Toshiaki Mitsui, Vidyasagar Pari, Vividha M. Sirsat, Ümit Haydar Erol

First published in London, United Kingdom, 2025 by IntechOpen
IntechOpen is the global imprint of INTECHOPEN LIMITED, registered in England and Wales, registration number: 11086078, 167-169 Great Portland Street, London, W1W 5PF, United Kingdom

For EU product safety concerns: IN TECH d.o.o., Prolaz Marije Krucifikse Kozulić 3, 51000 Rijeka, Croatia, info@intechopen.com or visit our website at intechopen.com.

British Library Cataloguing-in-Publication Data
A catalogue record for this book is available from the British Library

Advances in Plant Breeding - From Techniques to Stress Tolerance
Edited by Murat Aycan
p. cm.
Print ISBN 978-1-83635-313-3
Online ISBN 978-1-83635-312-6
eBook (PDF) ISBN 978-1-83635-314-0

Meet the editor

Dr. Murat Aycan is an Assistant Professor and an Academic Editor with expertise in plant molecular biology, stress physiology, and sustainable agriculture. He received his Ph.D. in Agronomy and has published extensively on abiotic stress tolerance, transgenerational inheritance, and the molecular mechanisms underlying responses to salinity and drought in crops. His research integrates genomics, transcriptomics, epigenetics, and microbiome studies to develop innovative breeding strategies. He has authored and co-edited several scientific works and has contributed to international collaborations on crop improvement. Currently based at the Laboratory of Biochemistry, Faculty of Agriculture, Niigata University, Japan, Dr. Aycan is dedicated to advancing plant breeding approaches to address global food security challenges.

Contents

Preface

Plant breeding has remained at the forefront of agricultural innovation, ensuring food security while addressing the mounting challenges of climate change, environmental stress, and a growing global population. This edited book, *Advances in Plant Breeding – From Techniques to Stress Tolerance*, brings together diverse perspectives on the latest breeding and molecular approaches aimed at developing stress-resilient and nutritionally enhanced crops.

The book begins with Chapter 1, *Advancements in Drought and Salt-Tolerant Crops: A Comparison of Breeding and Genetic Engineering Approaches*, which explores conventional and modern methods to mitigate major abiotic stresses that affect global crop productivity. Building on this, Chapter 2, *Molecular Breeding Strategies for Developing Salt Stress-Tolerant Rice*, focuses on one of the most vital crops in the world, presenting advanced approaches to tackle salinity stress, which poses a severe limitation to rice production in many regions.

Moving into biotic stress resistance, Chapter 3, *Plant Breeding for Tolerance to Biotic Stressors Including Nematodes, Diseases, and Insect Pests* provides a comprehensive overview of breeding strategies addressing a broad spectrum of biotic challenges. This is followed by Chapter 4, *Introgression of Biotic Stress Resistance Genes: Enhancing Crop Protection through Molecular Breeding*, which examines the integration of molecular tools for transferring resistance genes and achieving sustainable protection against pests and diseases.

Expanding the focus, Chapter 5, *Genetic Engineering of Agricultural Crops for Food and Nutritional Security*, discusses biotechnological strategies for enhancing food quality, nutrition, and overall crop resilience. Finally, Chapter 6, *Perspective Chapter: Seed Priming – A Novel Approach to Combat Microgravity Associated Stress Management in Plants*, highlights an innovative concept linking plant science to space biology.

Together, these chapters provide readers with a balanced coverage of abiotic and biotic stress tolerance, genetic engineering applications, and novel emerging approaches in plant breeding. The insights shared here will be valuable to researchers, students, and practitioners aiming to integrate traditional breeding with cutting-edge molecular and biotechnological tools for sustainable agriculture.

Dr. Murat Aycan
Laboratory of Biochemistry,
Faculty of Agriculture,
Niigata University,
Niigata, Japan

Section 1

Advances in Breeding for Abiotic Stress Tolerance

Chapter 1

Advancements in Drought and Salt-Tolerant Crops: A Comparison of Breeding and Genetic Engineering Approaches

Ritwik Acharya, Madhushri Das Datta,
Debnirmalya Gangopadhyay, Shubhajit Shaw, Rahul Chatterjee
and Ankita Manna

Abstract

Drought and salinity are major abiotic stresses threatening global agriculture, necessitating resilient crop development. This review compares traditional breeding and genetic engineering approaches to enhance stress tolerance. Traditional methods—such as selection, hybridization, and marker-assisted breeding—have achieved moderate gains but are limited by complex trait inheritance and lengthy breeding cycles. In contrast, genetic engineering enables precise and rapid improvements through transgenic expression of stress-responsive genes (e.g., DREB, NHX1, Late Embryogenesis Abundant (LEA)) and genome-editing tools like CRISPR/Cas9. Key physiological and molecular responses include stomatal regulation, ion homeostasis, osmotic adjustment, and antioxidative defense. Field successes like MON 87460 maize and Bt cotton illustrate the potential of biotechnological interventions. This review emphasizes integrative strategies combining molecular genetics, genomic selection, and high-throughput phenotyping for effective crop improvement. While genetic engineering holds significant promise, it faces challenges related to technical constraints, regulations, and socio-ethical concerns, especially in developing countries. The review advocates a synergistic model that blends the strengths of conventional breeding with gene-editing precision to accelerate the creation of climate-resilient crops. This concise synthesis supports innovation in plant biotechnology to sustain agricultural productivity under escalating environmental stress.

Keywords: drought tolerance, salt stress, genetic engineering, traditional breeding, CRISPR/Cas9, climate-resilient crops

1. Introduction

Climate change and increasing global food demands have underscored the urgent need for crops capable of withstanding challenging environmental conditions, particularly in arid and semi-arid regions where agricultural productivity is most vulnerable. Among the numerous abiotic stresses that threaten crop yields, drought and soil salinity are two of the most critical factors limiting agricultural output on a global scale [1, 2]. These stresses disrupt plant physiological and biochemical processes, ultimately reducing growth and productivity. Alarmingly, by 2050, it is projected that more than half of the world's arable land could be severely affected by water scarcity and salinity stress [3]. This looming crisis poses a major threat to global food security, particularly in regions already facing food shortages. As conventional agricultural practices struggle to keep pace with these mounting environmental pressures, there is an increasing need to develop resilient crop varieties to withstand such stresses to achieve sustainable agriculture and long-term food security [4]. **Figure 1** summarizes the impact of drought and salinity on stress-sensitive plants and highlights the superior performance of stress-tolerant plants under these adverse conditions.

To address these challenges, traditional plant breeding methods have been used to enhance crop tolerance to drought and salinity. These strategies have yielded significant improvements in some crops [5]. However, conventional breeding is often a slow and labor-intensive process, constrained by the limited genetic variability available within a species [6]. The emergence of genetic engineering and molecular biotechnology has revolutionized crop improvement strategies. These advanced tools enable the direct manipulation of specific genes associated with stress responses, allowing for the development of genetically modified crops with enhanced tolerance to drought and salinity [7]. Such innovations hold immense potential for future agriculture.

This article delves into the significant advancements achieved through both traditional breeding and modern genetic engineering strategies aimed at enhancing drought and salt tolerance in crops. By critically examining and comparing their underlying methodologies, key achievements, limitations, and potential for large-scale application, we seek to offer a well-rounded and in-depth perspective. The discussion highlights how these approaches contribute to improving crop resilience under abiotic stress conditions and evaluates their roles and future prospects in promoting sustainable and climate-resilient agriculture.

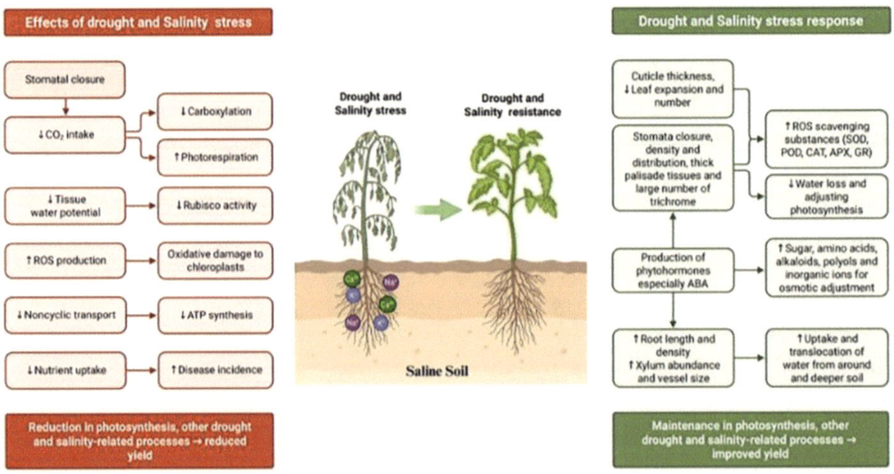

Figure 1.
Impact of drought and salinity on stress-sensitive plants and the superior performance of stress-tolerance plants under these adverse conditions.

DOI: http://dx.doi.org/10.5772/intechopen.1011113

2. Physiological and molecular mechanisms underlying plant tolerance to drought and salinity stresses

Plants have developed a sophisticated array of physiological and molecular strategies to survive drought and salinity, two major abiotic stresses that significantly constrain global agricultural productivity. These stresses disrupt water balance, ion homeostasis, and metabolic function, prompting plants to initiate multifaceted tolerance responses that span from whole-organism adjustments to molecular-level regulation.

At the physiological level, drought stress leads to a reduction in water availability, compelling plants to optimize water-use efficiency. A key mechanism involves the regulation of stomatal aperture through abscisic acid (ABA)-mediated signaling, which reduces transpiration-driven water loss [8]. Osmotic adjustment, achieved through the accumulation of compatible solutes such as proline, glycine betaine, and trehalose, helps maintain cellular turgor pressure and stabilizes proteins and membranes under water-deficient conditions [5]. The root system architecture also adapts to drought, with increased depth and density facilitating water uptake from deeper soil layers [9].

Under salt stress, plants face ionic toxicity and osmotic imbalance, primarily due to excessive sodium (Na^+) uptake. To combat this, selective ion transport mechanisms restrict Na^+ entry and sequester excess Na^+ into vacuoles *via* Na^+/H^+ antiporters, thereby preserving cytoplasmic ion homeostasis [2]. Maintaining a high potassium-to-sodium (K^+/Na^+) ratio is crucial for sustaining enzymatic activity and metabolic functions [10].

Both drought and salinity also trigger the overproduction of reactive oxygen species (ROS), such as superoxide radicals and hydrogen peroxide, which cause oxidative damage to cellular components. In response, plants activate enzymatic antioxidant defenses, including superoxide dismutase (SOD), catalase (CAT), and ascorbate peroxidase (APX), which detoxify ROS and mitigate oxidative stress [11]. Photosynthetic efficiency is safeguarded through non-photochemical quenching and chloroplast-based antioxidant systems that minimize photodamage and maintain energy production under stress [12].

Central to these physiological defenses is the ABA signaling pathway, which not only regulates stomatal closure and osmotic adjustment but also interacts closely with ROS detoxification mechanisms. This crosstalk enables plants to orchestrate a coordinated stress response, integrating hormonal signaling with redox regulation. Transcription factors such as DREB and MYB serve as crucial mediators in this network, regulating the expression of stress-responsive genes involved in osmoprotection, antioxidative defense, and ion homeostasis [2].

On the molecular level, stress perception at the cell membrane activates complex intracellular signaling cascades. Secondary messengers, such as calcium ions (Ca^{2+}), ROS, and phospholipids, initiate these responses by activating kinases, including mitogen-activated protein kinases (MAPKs) and calcium-dependent protein kinases (CDPKs), which in turn regulate downstream gene expression [11, 13]. Major transcription factor families—such as DREB, NAC, MYB, and bZIP—control the transcription of genes encoding osmoprotectants, ion transporters, and antioxidants [14]. Under drought, ABA biosynthesis and signaling genes, including NCED and AREB1, are upregulated to enhance tolerance through stomatal regulation and osmotic balance [15]. During salt stress, the Salt Overly Sensitive (SOS) pathway becomes crucial, with SOS1 functioning as a plasma membrane Na^+/H^+ antiporter that expels excess Na^+ and maintains ionic equilibrium [16].

Additionally, gene expression is modulated post-transcriptionally by small RNAs, particularly microRNAs (miRNAs), which fine-tune stress responses by targeting mRNAs for degradation or translational inhibition [17]. Epigenetic modifications, including DNA methylation and histone acetylation, contribute further to stress adaptation by regulating chromatin structure and enabling stress memory, thereby priming plants for enhanced responses to recurrent stresses [18].

Collectively, these multilayered physiological and molecular mechanisms equip plants with the resilience necessary to survive environmental adversities. Deciphering these complex networks offers critical insights for crop improvement strategies aimed at enhancing stress tolerance in the face of ongoing climate change.

3. Breeding approaches for developing tolerant crops

Abiotic stresses, such as salinity and drought, significantly constrain global agricultural productivity, threatening food security under changing climatic conditions [19]. Developing crop varieties with enhanced tolerance to these stresses is therefore a critical objective in plant breeding programs. Conventional breeding, marker-assisted selection, and transgenic approaches have been employed to incorporate stress-tolerance traits into high-yielding cultivars [20, 21].

3.1 Traditional breeding methods

Traditional breeding techniques have been pivotal in developing crop varieties resilient to environmental stresses. These methods, which include selection, hybridization, and mutation breeding, aim to enhance stress tolerance by leveraging natural genetic variability. Selection involves identifying and propagating plants with desirable traits, such as drought tolerance or disease resistance, through iterative cycles over generations, leading to improved stress-resilient varieties. Hybridization, by crossing genetically diverse parental lines, combines advantageous traits to produce offspring with superior stress tolerance and performance. Mutation breeding, which employs chemical or radiation-induced mutagenesis, facilitates the emergence of novel traits—including enhanced stress tolerance—that are subsequently integrated into breeding programs [22, 23]. Collectively, these approaches have played a crucial role in advancing the development of drought- and salt-tolerant crop varieties, enabling farmers to combat the detrimental effects of water scarcity and soil salinity [24, 25].

3.2 Marker-assisted selection (MAS) and quantitative trait loci (QTL) mapping

Quantitative traits, such as those related to drought and salt tolerance, are typically governed by a complex interplay of genetic and environmental factors. The genomic regions controlling these traits, known as quantitative trait loci (QTLs), are specific segments of the genome associated with particular traits. Advances in molecular genetics have greatly facilitated the identification of QTLs, initially driven by the development of linkage maps in the 1980s. These maps enabled the generation of mapping populations and the identification of polymorphic markers, establishing critical linkages between molecular markers and QTLs [26].

Molecular markers, categorized into protein-based markers (e.g., isozymes) and DNA-based markers (e.g., RFLP, AFLP, SSR, and SNP), have proven instrumental in marker-assisted selection (MAS). MAS accelerates the breeding process by selecting favorable alleles, improving complex traits such as drought-related yield and difficult-to-measure traits like root architecture, water-use efficiency, and osmotic adjustment. Moreover, these markers facilitate the introgression of desirable traits from wild relatives, reducing linkage drag by minimizing the unintended co-transfer of undesirable genes [27].

Recent advancements extend beyond marker-based approaches to include gene identification strategies within QTL regions. Traditional map-based cloning is now complemented by high-throughput technologies such as microarray-based transcriptional profiling, which elucidates differential gene expression under stress conditions [28, 29]. Integrative approaches combining genetic mapping with expression profiling have further refined the identification of candidate genes associated with stress tolerance [30, 31]. For example, Huang et al. [32] identified approximately 2000 drought-responsive genes in *Arabidopsis thaliana*, with a significant proportion regulated by abscisic acid (ABA). Similarly, Seki et al. [33] reported 277 drought-responsive genes, many of which also respond to cold and heat stress, highlighting the overlap in abiotic stress responses.

QTL studies in crops such as wheat, rice, cotton, oilseeds, and forage species have underscored the utility of MAS for improving drought and salinity tolerance. Advanced backcross QTL analysis has also been employed to assess donor introgressions within elite genetic backgrounds, enhancing the efficiency of breeding programs [34]. Additionally, QTL mapping studies have identified numerous loci involved in salt stress responses [31, 35]. These developments underscore the transformative potential of molecular markers and integrative genomic approaches in accelerating the development of stress-tolerant crops, paving the way for sustained agricultural productivity under challenging environmental conditions [36].

4. Genetic engineering approaches for enhancing drought and salinity tolerance in crops

Genetically modified (GM) crops have emerged as powerful tools to combat the increasing threat of abiotic stresses such as drought and salinity. These crops are developed using precise genetic engineering techniques that involve the insertion of foreign genes into plant genomes through methods like *Agrobacterium*-mediated transformation or direct gene transfer, leading to the creation of transgenic plants [37]. Unlike traditional breeding, which is limited by species compatibility, genetic engineering allows the introduction of genes across taxonomic boundaries, thereby accelerating trait enhancement [38].

A major focus of genetic modification is improving tolerance to complex, multigenic stresses such as salinity and drought. Salt stress activates several physiological responses in plants, including osmolyte accumulation, activation of antioxidant systems, and regulation of ion homeostasis. Incorporating salt-tolerance genes from halophytes or modulating native stress-responsive genes enhances ionic regulation and stress signaling pathways. Notably, chloroplast-targeted expression of osmolyte biosynthetic genes has shown superior outcomes, indicating that subcellular localization plays a pivotal role in transgene performance [39, 40]. Similarly, the introduction of genes related to drought tolerance, such as those encoding transcription factors, aquaporins, Late Embryogenesis Abundant (LEA) proteins, and enzymes involved in osmolyte biosynthesis, has improved water-use efficiency and conferred protection against cellular dehydration [41, 42]. Prominent examples include MON 87460 maize and transgenic rice expressing CSPB, Ubi1::TPSP, and NHX1, which have demonstrated enhanced tolerance under field conditions, supporting their agricultural viability.

Improving abiotic stress resilience is critical for ensuring global food security amid escalating climate change impacts [43]. Molecular breeding and biotechnology have focused on manipulating key transcription factors such as DREBs [44] and on leveraging LEA proteins, which stabilize macromolecules under stress due to their hydrophilic and thermostable nature [41]. Genetic modification of ion transporters like HKT1 and NHX1 has also proven vital for salinity tolerance by maintaining ionic homeostasis [45, 46]. Further, enhancing the expression of heat shock proteins (Hsps) has been shown to improve thermotolerance in several crops, as demonstrated by Bhatnagar-Mathur et al. [7].

CRISPR/Cas9 genome editing has recently revolutionized crop improvement by allowing precise modifications of stress-related genes [47, 48]. For example, DREB transcription factors confer stress tolerance, but constitutive overexpression may cause growth defects. Using stress-inducible or tissue-specific promoters allows for localized expression, mitigating growth penalties while maintaining stress resilience [40]. In soybean, overexpression of DREB1A improved drought response in greenhouse conditions through reduced water use, although field yields remained unchanged [44, 49].

LEA proteins, initially identified in seeds, have been recognized for their protective roles in vegetative tissues under abiotic stress. They are grouped into seven subfamilies and are highly expressed across plant genomes: 51 genes in *Arabidopsis*, 34 in rice, 108 in *Brassica napus*, and 23 in *Phyllostachys edulis* [50, 51]. Transgenic expression of wheat-derived group I and II LEA genes (PMA1959, PMA80) in rice enhanced drought and salinity tolerance [52], while OsEm1 upregulated other LEA genes under stress [53]. Group II proteins, such as RAB16A and SmLEA2, improved antioxidant defenses [54, 55], and group III members, such as HVA1 and OsG3LEA-47.3, imparted combined drought and heat tolerance [56, 57]. Additionally, atypical group V proteins, such as AdLEA from *Arachis diogoi*, enhanced multiple stress tolerances in tobacco [58].

Sodium transporters play a pivotal role in salinity stress responses. In rice, OsHKT1;1 restricts Na^+ accumulation in shoots and is regulated by the MYB-type transcription factor OsMYBc, which binds directly to its promoter. Mutants lacking OsMYBc show diminished OsHKT1;1 expression and increased salt sensitivity [45]. Similarly, vacuolar sequestration *via* NHX1 and cytoplasmic Na^+ efflux by SOS1 work synergistically to maintain Na^+/K^+ balance. Individually, these transporters are insufficient under high salinity (>200 mM NaCl), but co-expression of AtNHX1 and AtSOS1 in *Arabidopsis* allowed tolerance up to 250 mM NaCl with no yield loss [46].

The heat shock response is orchestrated by heat shock transcription factors (HSFs), which become activated upon stress and promote the expression of Hsps that stabilize proteins and form stress granules [7, 59–62]. This response is increasingly harnessed in transgenic plants to improve thermotolerance.

Salinity disrupts nutrient balance, membrane function, and photosynthesis due to elevated Na^+ and Cl^- [63–65], which plants counter through ion transporters and reactive oxygen species detoxification [2]. CRISPR/Cas9 offers a precise solution to these challenges. As illustrated in **Figure 2**, CRISPR/Cas9 enables targeted knockouts or overexpression of genes related to drought and salinity tolerance, allowing fine-tuned genetic interventions. Examples include overexpression of SOS1 and

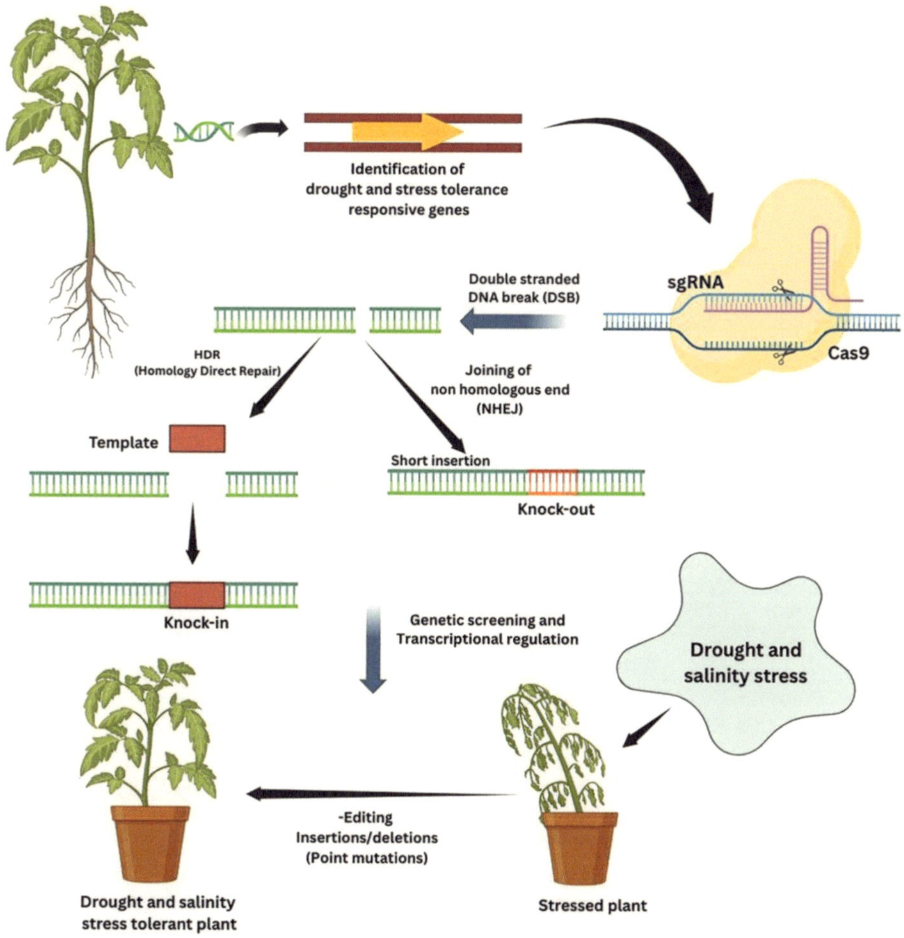

Figure 2.
The mechanism of CRISPR/Cas9-mediated development of drought and salt tolerance in plants.

HvHKT2;1 in *Arabidopsis* and barley for enhanced Na^+ transport [66, 67]. In rice, CRISPR knockouts of OsRR22, OsPQT3, and OsmiR535 improve salt stress tolerance [47, 68]. Similarly, editing SlHyPRP1 in tomato and ACQOS in *Arabidopsis* boosts stress resilience [48, 69]. Drought tolerance is polygenic and difficult to improve through traditional methods [70, 71]. CRISPR addresses this by modifying genes such as DERF1, MYB5, SPP, OST2, and miR169a in rice and *Arabidopsis* [72, 73], as well as ARGOS8, abh2, and ERA1 in cereals [74–77]. Successes in rapeseed, wheat, tomato, and cotton validate CRISPR's broad applicability in engineering climate-resilient crops [78, 79]. A summary of major CRISPR-mediated improvements in abiotic stress tolerance is presented in **Table 1**.

Further, pangenome analysis, gene stacking, and synthetic biology represent cutting-edge strategies in the development of drought- and salinity-tolerant plants. Pangenome analysis allows for the comprehensive exploration of genetic diversity across cultivated and wild relatives, facilitating the identification of novel alleles associated with osmotic adjustment, ion homeostasis, and stress signaling [95]. These unique alleles can then be targeted for introgression through marker-assisted back-crossing or genome editing to enhance stress resilience. Gene stacking, or pyramiding,

Plant	Targeted gene(s)	Gene function	Target trait	Reference
Rice	DERF1, PMS3, MSH1, MYB5, SPP	Involved in amino acid biosynthesis and regulation of drought-responsive genes; enhances tolerance via metabolic adaptation and stress-responsive transcription.	Drought tolerance	Zhang et al. [80].
Rice	ERA1	Negative regulator of abscisic acid (ABA) signaling; modulates dehydration stress response pathways.	Drought tolerance	Ogata et al. [77].
Rice	SRL1, SRL2	Control of leaf rolling by influencing leaf morphology, aiding in water conservation under drought.	Drought tolerance	Liao et al. [81].
Rice	SPL10	Regulates trichome (leaf hair) development, which can reduce water loss and improve salt stress adaptation.	Salinity tolerance	Lan et al. [82].
Rice	SOS1	Encodes a plasma membrane Na^+/H^+ antiporter; involved in sodium ion efflux, critical for salt tolerance.	Salinity tolerance	Lu et al. [83].
Tomato	HyPRP1	Hybrid proline-rich protein; acts as a transmembrane protein associated with multiple stress responses, especially salinity.	Salinity tolerance	Tran et al. [48].
Tomato	ABIG1	Homeodomain-leucine zipper (HD-ZIP) transcription factor involved in ABA-mediated stress response.	Salinity tolerance	Ding et al. [84].
Tomato	LBD40	Lateral organ boundary domain gene; involved in jasmonic acid (JA) mediated response to environmental stresses.	Drought tolerance	Liu et al. [85].
Tomato	NPR1	Encodes the receptor of salicylic acid; key regulator of systemic acquired resistance and drought stress responses.	Drought tolerance	Li et al. [86].
Wheat	DREB2, DREB3, ERF3	Transcription factors from DREB (dehydration-responsive element binding) and ERF families; modulate expression of drought-responsive genes.	Drought tolerance	Kim et al. [87].
Wheat	Two HAG homologs	Histone acetyltransferase-related genes; modulate reactive oxygen species (ROS) signaling and stress gene activation.	Salinity tolerance	Zheng et al. [88].
Maize	abh2	Abscisic acid 8′-hydroxylase; regulates stomatal opening and ABA catabolism, influencing drought response.	Drought tolerance	Liu et al. [89].
Maize	HKT1	High-affinity potassium transporter 1; regulates Na^+ exclusion from shoots and maintains ion homeostasis under salinity.	Salinity tolerance	Zhang et al. [90].
Cotton	AITR genes (e.g., DPA4, SOD7)	AITR family genes regulate plant architecture and seed size; modulate stress responses, including salt tolerance.	Salinity tolerance	Wang et al. [91].
Barley	HVP10	Involved in Na^+ sequestration in the tonoplast.	Salinity tolerance	Fu et al. [92].
Soybean	NHX5	Na^+/H^+ exchanger (NHX) transmembrane protein	Salinity tolerance	Sun et al. [93].
Pumpkin	RBOHD	NADPH oxidase	Salinity tolerance	Huang et al. [94].

Source: Compiled by the authors.

Table 1.
Significant CRISPR/Cas9 mediated drought and salinity tolerance enhancement.

involves the combination of multiple genes or quantitative trait loci (QTLs) that regulate different facets of stress response, such as osmoprotectant biosynthesis, antioxidative defense, and ion transport. This approach has been shown to produce crops with broader and more stable tolerance to multiple abiotic stresses [96]. In parallel, synthetic biology offers the potential to engineer entirely novel regulatory circuits and metabolic pathways not found in nature. For instance, the construction of synthetic osmoregulatory networks or stress-inducible gene modules in model plants has demonstrated enhanced adaptation under extreme environmental conditions [97]. While still largely in the experimental phase, these integrative technologies significantly expand the toolkit for developing climate-resilient crops and hold promise for future agricultural sustainability.

5. Practical outcomes of genetic engineering for drought and salt tolerance

Traditional breeding has long enhanced crop resilience by selecting tolerant cultivars through crossbreeding and exploiting natural genetic variability, as seen in India's development of salt- and drought-tolerant rice [98, 99]. However, this approach is time-intensive and limited by the gene pool's diversity. In contrast, genetic engineering enables precise and rapid modification of stress-related genes, exemplified by transgenic crops such as Bt cotton and drought-tolerant maize [49, 100]. Advances like CRISPR-Cas9 further allow targeted, foreign DNA-free edits, as demonstrated in drought- and salt-resistant tomato and rice [101]. While genetic engineering offers speed and specificity, traditional breeding garners broader acceptance. Integrating both strategies with genomics-assisted tools presents a robust pathway to sustainable agricultural resilience [102].

5.1 Successful GM and transgenic crops

The intensifying impacts of climate change, notably water scarcity and soil salinization, have propelled the development of genetically engineered crops with enhanced stress tolerance [103]. Biotechnological strategies aimed at increasing salt tolerance and water-use efficiency (WUE) are crucial for maintaining crop productivity under adverse conditions [104]. **Table 2** outlines various significant experiments on drought- and salt-tolerant crops. Drought-tolerant maize (MON 87460), engineered with the *Bacillus subtilis* cspB gene, enhances WUE by stabilizing proteins and photosynthetic structures under water deficit [117]. Bt cotton, initially designed for pest resistance, also improves drought tolerance *via* better stomatal regulation and root architecture [118]. Transgenic rice expressing AtNHX1 facilitates vacuolar Na^+ sequestration, maintaining ionic balance under salinity stress [119], while tomatoes overexpressing SlNCED1 increase abscisic acid, promoting osmotic balance and stomatal closure [120]. In cotton, transgenic stacking of HVA1 and DREB enhances osmotic adjustment and photosynthetic efficiency under drought, achieving up to 30% higher lint yields while maintaining fiber quality [121–123]. These traits demonstrate the value of combining biotic and abiotic resistance [124], though biosafety, regulatory approval, and public acceptance remain essential [125]. CRISPR/Cas9-based editing of ARGOS8 in maize reduces ethylene sensitivity, delaying senescence and boosting biomass without yield penalties [126, 127], exemplifying integrated molecular strategies for climate-resilient agriculture [128].

Crop	Genotypes studied	Number of genotypes	Nature of salinity stressor	Nature of study	Parameters studied	Trait targeted	Method used	Key outcomes	Reference
Bt Cotton	Bt cotton hybrids	3	Salinity	Field trials across India	Yield, pest resistance	Salt tolerance and pest resistance	Genetic modification (Bt gene insertion)	Bt cotton showed increased yield and reduced pesticide use under saline conditions.	Qaim [105].
Drought-resistant maize	33D53AM, conventional hybrid; PAM, DT hybrid	2	Drought and salinity	Regional-scale simulation study	Yield, water-use efficiency	Drought and salinity tolerance	APSIM-Maize modeling	DT hybrids showed improved yield and water savings under deficit irrigation.	Su et al. [106].
Transgenic rice	Various transgenic lines	Multiple	Drought and salinity	Greenhouse and field trials	Growth parameters, ion content, and yield	Salinity tolerance	Genetic transformation	Transgenic lines exhibited improved growth and yield under saline conditions.	Ashraf and Akram [107], Joshi et al. [108]
Salt-tolerant *Acacia*	Various Acacia species	Multiple	Salinity	Field trials	Survival rate, growth parameters	Salinity tolerance	Field evaluation	Identified species with higher survival and growth under saline conditions.	Niknam and McComb [109].
Barley	BH 19-15, BH 19-49, BH 19-02, BH 946, BH 20-02, RD 2794, BH 20-36, BH 19-52, BH 20-38, BH 19-44, BH 20-40, BH 20-09, BH 19-13, DWRB 91	14	Natural soil salinity (EC 4 dS/m)	Field trial	Yield components, stress indices (SSI, TOL, STI, etc.)	Salinity tolerance	Field evaluation with stress indices	Identified genotypes with superior performance under salinity stress.	Kumar et al. [110].
Tomato	Various cultivars	Multiple	Drought and salinity	Controlled environment	Growth parameters, ion content, and yield	Salinity and drought tolerance	Physiological and agronomic assessments	Certain cultivars maintained better growthand yield under salinity stress.	Maryum et al. [111], Murtaza et al. [112].

Crop	Genotypes studied	Number of genotypes	Nature of salinity stressor	Nature of study	Parameters studied	Trait targeted	Method used	Key outcomes	Reference
Wheat	Kharchia-65, KRL-210, HD-2329, WH-542	4	200 mM NaCl	Controlled environment	Plant height, tiller number, leaf senescence, and chlorophyll content	Salinity tolerance	Physiological and biochemical analyses	Kharchia-65 and KRL-210 exhibited higher tolerance with less reduction in growth parameters.	Kumar et al. [113].
Wheat	GW503, DBW17,NI5643, NW1014, PBW65, PBW502, DBW187, DBW222, DBW303, NW1076, HD1941, HD2009, HD3086, GW89, K9162	20	EC 4.02 dS/m (saline-sodic field conditions)	Field trial	Yield components, ion content, proline accumulation, and antioxidant activity	Salinity tolerance	Field evaluation with biochemical assays	Identified tolerant genotypes with higher antioxidant activity and better ion balance.	Patwa et al. [114].
Rice	HKN, XD2H, HHZ, DJWJ, JFX, NSIC Rc294	6	Saline field conditions	Field trial	Yield and yield components	Salinity tolerance	Agronomic evaluation	Tolerant genotypes maintained higher yield under salinity stress.	Xu et al. [115].
Cotton	Z9807, Z0228, Z7526,Z0710, Z7514, Z1910, Z7516,Z0102, Z7780, Z9648, Z9612	11	Salinity	Seedling stage evaluation	Sodium and potassium ion content, salt tolerance index	Salt tolerance	Morphological and physiological assessments	Identified genotypes with higher salt tolerance based on ion content and growth parameters.	Sikder et al. [116].

Source: Compiled by the authors.

Table 2.
Significant transgenic experiments to develop drought and salinity tolerance in different crop varieties.

5.2 Field performance and adoption of genetically modified (GM) crops

Genetically modified (GM) crops offer several agronomic and nutritional benefits, including herbicide tolerance, insect resistance, abiotic stress tolerance, disease resistance, and enhanced nutritional content [38]. Bt cotton, a leading example, has significantly reduced pesticide use and improved yields and profitability in countries like China and India [129, 130]. Similarly, Bt soybean and Bt maize have achieved effective pest control in North America [131]. GM crops with herbicide tolerance, such as glyphosate-resistant maize and soybean, lower labor costs and soil degradation by minimizing tillage [132]. Nutritionally enhanced GM crops, such as Golden Rice, address deficiencies such as vitamin A shortage [133]. However, public skepticism and regulatory constraints—especially stringent in Europe—hinder broader adoption [134, 135]. While GM crops show exceptional field performance in North and South America, concerns over resistance development, gene flow, and biodiversity impacts continue to shape their global acceptance and future deployment [136].

6. Comparing breeding and genetic engineering approaches

Traditional plant breeding and genetic engineering are two pivotal strategies for enhancing drought and salinity tolerance in crops. Conventional breeding involves selecting and crossing naturally stress-tolerant varieties to combine beneficial traits. This method has yielded notable successes, such as the development of salt-tolerant rice and wheat lines. However, its effectiveness is limited by the availability of genetic diversity, sexual compatibility barriers, and the complexity of traits such as drought and salinity tolerance, which are often controlled by multiple genes [137, 138]. Moreover, the long breeding cycles and potential for linkage drag make this approach time-consuming and less precise.

In contrast, genetic engineering enables the direct introduction or modification of specific genes associated with stress tolerance, bypassing species barriers. Key

Criteria	Traditional breeding	Genetic engineering	Reference
Definition	Crossing of organisms with desirable traits through sexual reproduction.	Direct manipulation of an organism's DNA to introduce or alter traits.	Acquaah [140], Nicholl [141].
Speed	Slow – often requires multiple generations over several years.	Fast–specific traits can be introduced within weeks or months.	Acquaah [140], Nicholl [141].
Precision	Low – involves large segments of DNA, and unintended traits may be inherited.	High – allows targeting of specific genes with minimal unintended effects.	Acquaah [140], Ladics et al. [142].
Gene sources	Limited to sexually compatible species.	Genes can come from any species (e.g., bacteria, animals, etc.).	Acquaah [140], Nicholl [141].
Trait predictability	Less predictable – influenced by recombination and environment.	More predictable – genes are selected and controlled more precisely.	Acquaah [140], Nicholl [141].

Criteria	Traditional breeding	Genetic engineering	Reference
Cost	Lower upfront costs but higher cumulative costs due to time.	Higher R&D and regulatory costs but more efficient in the long run.	Acquaah [140], National Academies of Sciences [143].
Regulatory oversight	Minimal – often exempt from modern biotech regulations.	Strict – requires comprehensive biosafety, health, and environmental evaluations.	Acquaah [140], Eckerstorfer et al. [144].
Public perception	Generally favorable – seen as "natural" or traditional.	Often controversial – concerns about GMOs and unnatural modifications.	Acquaah [140], Mueller and Flachs [145].
Notable examples	Hybrid corn, drought-tolerant wheat, improved tomatoes via selection.	Bt corn, Golden Rice (Vitamin A), herbicide-resistant soybeans, insulin from GM bacteria.	Rosero et al. [146], Bagwan et al. [147].
Limitations	Slower progress, limited gene pool, unpredictable outcomes.	Ethical debates, high regulation, potential for unintended ecological or health impacts.	Acquaah [140], Singer et al. [148].

Source: Compiled by the authors.

Table 3.
Outlines the key distinctions between traditional breeding and genetic engineering approaches in developing drought and salinity-stress-tolerant plants.

advances include the incorporation of genes encoding transcription factors such as DREB, Late Embryogenesis Abundant (LEA) proteins, and ion transporters like NHX1 and HKT1, which have demonstrated enhanced osmotic adjustment, ion homeostasis, and cellular protection under stress conditions [41, 44–46]. Moreover, using stress-inducible and tissue-specific promoters helps mitigate adverse effects on plant growth by restricting transgene expression to relevant conditions or tissues [139].

While breeding remains vital for integrating multiple traits into elite germplasm, genetic engineering offers complementary precision and speed essential for tackling complex, multigenic stress responses in a changing climate. **Table 3** outlines the key distinctions between traditional breeding and genetic engineering approaches in the development of stress-tolerant plants.

7. Integrating breeding and genetic engineering

The development of stress-resilient crops requires a comprehensive strategy that integrates conventional breeding with advanced genetic engineering methodologies. As illustrated in **Table 4**, several examples highlight the synergistic use of these approaches in developing drought- and salinity-tolerant plant varieties. Conventional breeding techniques, such as hybridization and selection, continue to play a pivotal role in exploiting naturally occurring genetic diversity within elite germplasm. Nevertheless, the efficiency of these methods is constrained by prolonged breeding cycles and the intricate genetic architecture of abiotic stress tolerance, particularly due to the multigenic nature of traits like drought and salinity resistance [103, 162].

Recent advances in genome editing, particularly CRISPR/Cas9, have enabled precise, targeted modifications of stress-responsive genes involved in ionic homeostasis

Strategy/Approach	Description	Examples	Advantages	Limitations	References
Conventional breeding	Selection and hybridization of drought/ salt-tolerant genotypes over generations.	Drought-tolerant wheat, salt-tolerant rice landraces.	Widely accepted; uses natural diversity.	Time-consuming; limited gene pool; complex traits hard to stabilize.	Collins et al. [149], Munns and Gilliham [150].
Marker-assisted selection (MAS)	Use of molecular markers linked to QTLs for stress tolerance.	QTLs for salt tolerance in rice (Saltol), drought QTLs in maize.	Speeds up breeding; improves precision.	Still constrained by natural variation; expensive for polygenic traits.	Raj and Nadarajah [151].
Transgenic approaches	Introduction of stress-tolerance genes (e.g., transcription factors, transporters).	DREB1A, AtNHX1, HKT1, P5CS in rice, tomato, and maize.	Enables introgression of novel genes; stress-specific expression.	Regulatory hurdles; public concerns; off-target effects.	Roy et al. [152], Zhang and Blumwald, [153], Bhatnagar-Mathur et al. [7].
CRISPR/Cas9 genome editing	Precise editing of native stress-response genes without introducing foreign DNA.	Editing OsRR22 for salt tolerance; ARGOS8 in maize for drought.	High precision; avoids transgenic classification in some countries.	Regulatory ambiguity; potential off-targets.	Shi et al. [76], Zhang et al. [47], Waltz [154].
Gene pyramiding	Stacking of multiple genes/ QTLs conferring different aspects of tolerance.	Combining DREB, LEA, and ion transporter genes.	Broader, more durable stress tolerance.	Complexity in regulation and expression balance.	Tester and Langridge [155].
Speed breeding + CRISPR	Rapid generation cycles with simultaneous genome editing.	Used in wheat and tomato to accelerate CRISPR-based trait introgression.	Significantly shortens time from lab to field.	Requires infrastructure; currently optimized for select crops.	Watson et al. [156], Hickey et al. [157].
Phenotyping and genomics integration	Use of high-throughput phenotyping + genome-wide association studies (GWAS).	Identifying novel alleles for osmotic stress resilience in rice.	Precision trait mapping; enables genotype-phenotype linkage.	High technical and data demands; phenotyping under field stress is difficult.	Varshney et al. [158], Tardieu et al. [159].
Synthetic biology approaches	Designing novel biosynthetic pathways or regulatory networks for stress resilience.	Engineering synthetic osmoregulatory circuits in model plants.	Expands beyond natural gene limitations.	Still largely experimental; complex regulatory approval.	Liu and Stewart [160], Schaumberg et al. [161].

Source: Compiled by the authors.

Table 4.
An account of the prominent studies on breeding and genetic engineering together for drought and salinity tolerance in crops.

and osmoprotectant synthesis, facilitating the rapid development of superior allelic variants [163]. CRISPR-edited rice and maize lines have demonstrated enhanced physiological responses under stress, including up to 30% improved water-use efficiency and a 25% reduction in sodium accumulation compared to non-edited counterparts [50, 114].

Integrating genetic engineering with hybrid breeding is further strengthened by high-throughput phenotyping and genomic prediction models, which help assess allele performance across diverse genotypes [164]. Pangenomic analyses of wild relatives provide access to untapped alleles for traits like osmotic adjustment, which can be fine-mapped and introgressed *via* marker-assisted backcrossing or targeted editing [95].

Collaborative platforms uniting public and private sectors enable shared phenotyping protocols and germplasm development, while machine learning algorithms optimize hybrid × gene-editing strategies. Together, these integrated methodologies have boosted drought tolerance gains by up to 40% in key crops such as wheat and cotton [165].

8. Challenges and considerations in the deployment of genetically modified crops

The development of genetically modified (GM) crops with enhanced tolerance to drought and salinity represents a promising strategy to strengthen global food security amid intensifying environmental stresses. Engineered through sophisticated molecular technologies, these crops offer precise interventions to improve plant resilience under adverse conditions. However, their widespread adoption is accompanied by several ecological, regulatory, and socio-economic challenges. A major ecological concern involves the potential for transgene flow to wild relatives, which may unintentionally introduce novel traits into natural ecosystems, thereby disrupting ecological balance and affecting non-target organisms [166, 167]. Furthermore, interactions between GM crops and various biotic components—such as pests, pollinators, and beneficial soil microbes—require rigorous ecological risk assessments to evaluate long-term environmental impacts. It is also critical to recognize the disparity between laboratory success and real-world performance, as many transgenic lines that exhibit stress tolerance under controlled conditions encounter limitations in field environments due to complex environmental interactions and regulatory constraints. Several notable examples illustrating these challenges are summarized in **Table 5**.

On the regulatory front, global frameworks remain fragmented. Countries differ significantly in how they classify and manage GMOs, particularly with the advent of gene-editing technologies such as CRISPR-Cas9. In the United States, a relatively permissive regulatory approach distinguishes between transgenic organisms and gene-edited crops lacking foreign DNA, often exempting the latter from stringent oversight and thereby facilitating their commercial release [174]. Conversely, the European Union adopts a more cautious stance. The European Court of Justice ruled in 2018 that all gene-edited organisms fall under existing GMO legislation, irrespective of the presence of foreign DNA [175]. In contrast, countries such as Argentina, Brazil, and Japan have adopted more nuanced frameworks, exempting CRISPR-edited crops from GMO regulations when modifications mimic natural mutations and no transgenes are introduced [176].

Beyond regulation, social acceptance remains a challenge. Despite scientific consensus on GM crop safety, public skepticism persists due to perceived health

Crop	Transgene(s)	Laboratory Outcomes	Field Performance	Reference
Wheat (HB4)	HaHB4 (from sunflower)	Enhanced drought tolerance; delayed senescence	Approved in Argentina (2020), Brazil (2021), USA (2022); increased yield under drought	Ribichich et al. [168], Raineri et al. [169].
Soybean (HB4)	HaHB4 (from sunflower)	Improved drought tolerance	Four percent average yield increase; up to 10.5% under high temperature and drought; approved in China (2022)	Ribichich et al. [168].
Soybean	AhBADH (from *Atriplex hortensis*)	Enhanced salt tolerance; stable gene expression	Lines TL2 and TL7 showed improved agronomic traits under 300 mM NaCl; under biosafety assessment	Yu et al. [170].
Wheat	MDAR1 (from Arabidopsis)	Improved salt stress tolerance	Field trials in Egypt demonstrated enhanced performance under salinity stress	Abdelsattar et al. [171].
Tomato	Overexpression of the salt-tolerance gene	Thrived in soils 50 times saltier than normal	Not yet commercially viable; projected 3+ years to market	Guo et al. [172].
Wheat (Kharchia 65)	Traditional breeding (Kharchia Local × rust-resistant lines)	High salt tolerance	Used as a standard for salt tolerance in wheat; yields 10–20 Q/ha	Sathee et al. [173].

Source: Compiled by the authors.

Table 5.
An account of laboratory success and practical field application, where many transgenic crops face environmental and regulatory screening.

risks, environmental impacts, and ethical concerns [134]. Addressing these issues requires transparent governance, robust public engagement, and science-based communication strategies. Economically, the high costs associated with developing GM crops—including investment in gene-editing tools, rigorous testing, regulatory approvals, and necessary infrastructure—can restrict access for smallholder farmers, particularly in developing regions [177]. This economic barrier risks exacerbating existing inequalities in agricultural productivity and technological adoption [178].

Thus, while GM crops and genome-edited technologies hold substantial potential to enhance crop resilience and agricultural sustainability, their broader success hinges on the implementation of integrative strategies that balance innovation with ecological safety, regulatory coherence, equitable access, and public trust.

9. Future prospects

The escalating impacts of climate change and the need to sustain a growing global population have highlighted the urgency of developing drought- and salt-tolerant crops. While traditional breeding has contributed to crop resilience, its slow pace and difficulty in combining multiple traits limit its effectiveness. The integration of biotechnology, particularly genetic engineering, has significantly enhanced breeding precision by enabling the targeted incorporation of genes associated with drought and salt tolerance [179]. Innovations such as CRISPR-Cas9 have further refined genome editing, allowing for precise trait introduction with minimal off-target effects compared to conventional methods [71]. These advances are pivotal in addressing climate-induced challenges and ensuring food security. Additionally, precision agriculture—combining biotechnology with data-driven tools such as remote sensing and soil moisture sensors—enables site-specific crop management, improving yield and sustainability [179]. The rapid development of gene-editing technologies and gene synthesis platforms also facilitates the design of stress-resilient crops tailored to local environments, ushering in an era of personalized agriculture [180]. Together, these integrated approaches offer transformative potential for cultivating climate-resilient crops and securing global food systems in an uncertain environmental future.

10. Conclusion

The escalating impacts of climate change, including intensified drought and soil salinization, present significant threats to agricultural sustainability and global food security. This review examines the physiological, molecular, and genetic mechanisms underlying plant responses to abiotic stress and assesses the effectiveness of both conventional breeding and modern genetic engineering in developing drought- and salt-tolerant crops. Traditional breeding, though widely utilized, is hindered by the slow integration of traits, polygenic complexity, and limited genetic diversity. In contrast, genetic engineering—particularly transgenic approaches and CRISPR/Cas9 genome editing—has enabled precise modification of key stress-responsive genes, enhancing water-use efficiency, ion regulation, and antioxidative responses. Successful examples include drought-tolerant maize (MON 87460) and salinity-tolerant rice expressing AtNHX1. However, the broader adoption of genetically modified crops remains challenged by technical, regulatory, and ethical concerns, such as rigorous biosafety protocols, public resistance, and limited access in low-resource regions. The review underscores the promise of integrative strategies combining traditional and molecular techniques, supported by genomics, high-throughput phenotyping, and predictive modeling, to accelerate breeding outcomes. As climate pressures intensify, future efforts must prioritize interdisciplinary collaboration, transparent regulation, and equitable innovation. Ultimately, a harmonized approach merging genetic precision with conventional resilience offers the most sustainable path toward climate-resilient agriculture.

Acknowledgements

The authors are grateful to the Hon'ble Vice Chancellor of Raiganj University and the Hon'ble Principal, Hiralal Mazumdar College for Women, India, for providing the necessary facilities. This research did not receive any specific grant from funding agencies in the public, commercial, or not-for-profit sectors.

Conflict of interest

The authors declare no conflict of interest.

Author details

Ritwik Acharya[1,2]*, Madhushri Das Datta[2], Debnirmalya Gangopadhyay[1], Shubhajit Shaw[1], Rahul Chatterjee[3] and Ankita Manna[4]

1 Department of Sericulture, Raiganj University, Raiganj, West Bengal, India

2 Department of Botany, Hiralal Mazumdar Memorial College for Women, Kolkata, West Bengal, India

3 Department of Environmental Science, Indira Gandhi National Open University (Study Centre), Howrah, West Bengal, India

4 Department of Botany, University of Calcutta (Jogamaya Devi College Campus), Kolkata, West Bengal, India

*Address all correspondence to: rittik.acharya@gmail.com

References

[1] Boyer JS. Plant productivity and environment. Science. 1982;**218**(4571):443-448

[2] Munns R, Tester M. Mechanisms of salinity tolerance. Annual Review of Plant Biology. 2008;**59**(1):651-681

[3] Khondoker M, Mandal S, Gurav R, Hwang S. Freshwater shortage, salinity increase, and global food production: A need for sustainable irrigation water desalination—A scoping review. Earth. 4 Apr 2023;**4**(2):223-240

[4] Lobell DB, Field CB. Global scale climate–crop yield relationships and the impacts of recent warming. Environmental Research Letters. 2007;**2**(1):014002

[5] Ashraf MF, Foolad MR. Roles of glycine betaine and proline in improving plant abiotic stress resistance. Environmental and Experimental Botany. 2007;**59**(2):206-216

[6] Tuberosa R, Salvi S. Genomics-based approaches to improve drought tolerance of crops. Trends in Plant Science. 2006;**11**(8):405-412

[7] Bhatnagar-Mathur P, Vadez V, Sharma KK. Transgenic approaches for abiotic stress tolerance in plants: Retrospect and prospects. Plant Cell Reports. 2008;**27**:411-424

[8] Schroeder JI, Allen GJ, Hugouvieux V, Kwak JM, Waner D. Guard cell signal transduction. Annual Review of Plant Biology. 2001;**52**(1):627-658

[9] Uga Y, Sugimoto K, Ogawa S, Rane J, Ishitani M, Hara N, et al. Control of root system architecture by deeper rooting 1 increases rice yield under drought conditions. Nature Genetics. 2013;**45**(9):1097-1102

[10] Shabala S, Cuin TA. Potassium transport and plant salt tolerance. Physiologia Plantarum. 2008;**133**(4):651-669

[11] Mittler R, Zandalinas SI, Fichman Y, Van Breusegem F. Reactive oxygen species signalling in plant stress responses. Nature Reviews Molecular Cell Biology. 2022;**23**(10):663-679

[12] Flexas J, Medrano H. Drought-inhibition of photosynthesis in C3 plants: Stomatal and non-stomatal limitations revisited. Annals of Botany. 2002;**89**(2):183-189

[13] Sinha A, Berkelhammer M, Stott L, Mudelsee M, Cheng H, Biswas J. The leading mode of Indian summer monsoon precipitation variability during the last millennium. Geophysical Research Letters. 2011;**38**(15):1-5

[14] Nakashima K, Yamaguchi-Shinozaki K, Shinozaki K. The transcriptional regulatory network in the drought response and its crosstalk in abiotic stress responses including drought, cold, and heat. Frontiers in Plant Science. 2014;**5**:170

[15] Cutler SR, Rodriguez PL, Finkelstein RR, Abrams SR. Abscisic acid: Emergence of a core signaling network. Annual Review of Plant Biology. 2010;**61**(1):651-679

[16] Zhu JK. Abiotic stress signaling and responses in plants. Cell. 2016;**167**(2):313-324

[17] Song J, Price DJ, Guvenen F, Bloom N, Von Wachter T. Firming up

inequality. The Quarterly Journal of Economics. 2019;**134**(1):1-50

[18] Kim KH, Kabir E, Kabir S. A review on the human health impact of airborne particulate matter. Environment International. 2015;**74**:136-143

[19] Al-Tawaha AR, Odat N, Benkeblia N, Kerkoub N, Labidi Z, Boumendjel M, et al. Breeding crops for tolerance to salinity, heat, and drought. Climate Change and Agriculture: Perspectives, Sustainability and Resilience. 18 Nov 2022:95-110

[20] Ashraf M, Foolad MR. Crop breeding for salt tolerance in the era of molecular markers and marker-assisted selection. Plant Breeding. 2013;**132**(1):10-20

[21] Kumar M, Prusty MR, Pandey MK, Singh PK, Bohra A, Guo B, et al. Application of CRISPR/Cas9-mediated gene editing for abiotic stress management in crop plants. Frontiers in Plant Science. 2023;**14**:1157678

[22] Dwivedi SL, Ceccarelli S, Blair MW, Upadhyaya HD, Are AK, Ortiz R. Landrace germplasm for improving yield and abiotic stress adaptation. Trends in Plant Science. 2016;**21**(1):31-42

[23] Zafar SA, Zaidi SS, Gaba Y, Singla-Pareek SL, Dhankher OP, Li X, et al. Engineering abiotic stress tolerance via CRISPR/Cas-mediated genome editing. Journal of Experimental Botany. 2020;**71**(2):470-479

[24] Prasanna BM, Hossain F, Lal R. Climate-resilient agriculture: Principles, approaches and challenges. Journal of Experimental Botany. 2020;**71**(15):4517-4525

[25] Hafeez U, Ali M, Hassan SM, Akram MA, Zafar A. Advances in breeding and engineering climate-resilient crops: A comprehensive review. International Journal of Research and Advances in Agricultural Sciences. 2023;**2**(2):85-99

[26] Venuprasad R, Dalid CO, Del Valle M, Zhao D, Espiritu M, Sta Cruz MT, et al. Identification and characterization of large-effect quantitative trait loci for grain yield under lowland drought stress in rice using bulk-segregant analysis. Theoretical and Applied Genetics. 2009;**120**:177-190

[27] Sanghvi GV, Dave GS. Molecular markers in plant biotechnology. In: Plant Biotechnology. Vol. 1. Florida: Apple Academic Press; 2017. pp. 415-453

[28] Sahi C, Singh A, Kumar K, Blumwald E, Grover A. Salt stress response in rice: Genetics, molecular biology, and comparative genomics. Functional & Integrative Genomics. 2006;**6**:263-284

[29] Walia H, Wilson C, Zeng L, Ismail AM, Condamine P, Close TJ. Genome-wide transcriptional analysis of salinity stressed japonica and indica rice genotypes during panicle initiation stage. Plant Molecular Biology. 2007;**63**:609-623

[30] Marino R, Ponnaiah M, Krajewski P, Frova C, Gianfranceschi L, Pè ME, et al. Addressing drought tolerance in maize by transcriptional profiling and mapping. Molecular Genetics and Genomics. 2009;**281**:163-179

[31] Pandit A, Rai V, Bal S, Sinha S, Kumar V, Chauhan M, et al. Combining QTL mapping and transcriptome profiling of bulked RILs for identification of functional polymorphism for salt tolerance genes in rice (*Oryza sativa* L.). Molecular Genetics and Genomics. 2010;**284**:121-136

[32] Huang D, Wu W, Abrams SR, Cutler AJ. The relationship of drought-related gene expression in *Arabidopsis thaliana* to hormonal and environmental factors. Journal of Experimental Botany. 2008;**59**(11):2991-3007

[33] Seki M, Narusaka M, Ishida J, Nanjo T, Fujita M, Oono Y, et al. Monitoring the expression profiles of 7000 Arabidopsis genes under drought, cold and high-salinity stresses using a full-length cDNA microarray. The Plant Journal. 2002;**31**(3):279-292

[34] Tanksley SD, Nelson JC. Advanced backcross QTL analysis: A method for the simultaneous discovery and transfer of valuable QTLs from unadapted germplasm into elite breeding lines. Theoretical and Applied Genetics. 1996;**92**:191-203

[35] Ammar MH, Pandit A, Singh RK, Sameena S, Chauhan MS, Singh AK, et al. Mapping of QTLs controlling Na+, K+ and CI− ion concentrations in salt tolerant indica rice variety CSR27. Journal of Plant Biochemistry and Biotechnology. 2009;**18**:139-150

[36] Rauf S, Al-Khayri JM, Zaharieva M, Monneveux P, Khalil F. Breeding strategies to enhance drought tolerance in crops. In: Advances in Plant Breeding Strategies: Agronomic, Abiotic and Biotic Stress Traits. Cham: Springer; 2016. pp. 397-445

[37] Griffiths AJ, Wessler SR, Lewontin RC, Gelbart WM, Suzuki DT, Miller JH. Introduction to Genetic Analysis. 8th (Ed.). New York [Internet]: Freeman WH; 2005

[38] Kumar K, Gambhir G, Dass A, Tripathi AK, Singh A, Jha AK, et al. Genetically modified crops: Current status and future prospects. Planta. 2020;**251**(4):91

[39] Ashraf M, Athar HR, Harris PJ, Kwon TR. Some prospective strategies for improving crop salt tolerance. Advances in Agronomy. 2008;**97**:45-110

[40] Fageria NK, Stone LF, Santos AB. Breeding for salinity tolerance. In: Plant Breeding for Abiotic Stress Tolerance. Berlin, Heidelberg: Springer; 2012. pp. 103-122

[41] Banerjee A, Roychoudhury A. Group II late embryogenesis abundant (LEA) proteins: Structural and functional aspects in plant abiotic stress. Plant Growth Regulation. 2016;**79**:1-7

[42] Haghpanah M, Hashemipetroudi S, Arzani A, Araniti F. Drought tolerance in plants: Physiological and molecular responses. Plants. 2024;**13**(21):2962

[43] KhokharVoytas A, Shahbaz M, Maqsood MF, Zulfiqar U, Naz N, Iqbal UZ, et al. Genetic modification strategies for enhancing plant resilience to abiotic stresses in the context of climate change. Functional & Integrative Genomics. 2023;**23**(3):283

[44] de Paiva Rolla AA, de Fátima Corrêa Carvalho J, Fuganti-Pagliarini R, Engels C, Do Rio A, Marin SR, et al. Phenotyping soybean plants transformed with rd29A: AtDREB1A for drought tolerance in the greenhouse and field. Transgenic Research. 2014;**23**:75-87

[45] Dave A, Agarwal P, Agarwal PK. Mechanism of high affinity potassium transporter (HKT) towards improved crop productivity in saline agricultural lands. 3 Biotech. 2022;**12**(2):51

[46] Paul S, Roychoudhury A. Transgenic plants for improved salinity and drought tolerance. In: Biotechnologies of Crop Improvement, Volume 2: Transgenic Approaches. Cham: Springer; 2018. pp. 141-181

[47] Zhang A, Liu Y, Wang F, Li T, Chen Z, Kong D, et al. Enhanced rice salinity tolerance via CRISPR/Cas9-targeted mutagenesis of the OsRR22 gene. Molecular Breeding. 2019;**39**:47

[48] Tran MT, Doan DTH, Kim J, Song YJ, Sung YW, Das S, et al. CRISPR/Cas9-based precise excision of SlHyPRP1 domain (s) to obtain salt stress-tolerant tomato. Plant Cell Reports. 2021;**40**:999-1011

[49] Liang C. Genetically modified crops with drought tolerance: Achievements, challenges, and perspectives. In: Drought Stress Tolerance in Plants, Vol. 2: Molecular and Genetic Perspectives. Cham: Springer; 2016. pp. 531-547

[50] Huang Z, Zhong XJ, He J, Jin SH, Guo HD, Yu XF, et al. Genome-wide identification, characterization, and stress-responsive expression profiling of genes encoding LEA (late embryogenesis abundant) proteins in Moso bamboo (*Phyllostachys edulis*). PLoS One. 2016;**11**(11):e0165953

[51] Liang Y, Xiong Z, Zheng J, Xu D, Zhu Z, Xiang J, et al. Genome-wide identification, structural analysis and new insights into late embryogenesis abundant (LEA) gene family formation pattern in *Brassica napus*. Scientific Reports. 2016;**6**(1):24265

[52] Cheng Z, Targolli J, Huang X, Wu R. Wheat LEA genes, PMA80 and PMA1959, enhance dehydration tolerance of transgenic rice (*Oryza sativa* L.). Molecular Breeding. 2002;**10**:71-82

[53] Yu J, Lai Y, Wu X, Wu G, Guo C. Overexpression of OsEm1 encoding a group I LEA protein confers enhanced drought tolerance in rice. Biochemical and Biophysical Research Communications. 2016;**478**(2):703-709

[54] Ganguly M, Datta K, Roychoudhury A, Gayen D, Sengupta DN, Datta SK. Overexpression of Rab16A gene in indica rice variety for generating enhanced salt tolerance. Plant Signaling & Behavior. 2012;7(4):502-509

[55] Wang H, Wu Y, Yang X, Guo X, Cao X. SmLEA2, a gene for late embryogenesis abundant protein isolated from salvia miltiorrhiza, confers tolerance to drought and salt stress in *Escherichia coli* and *S. Miltiorrhiza*. Protoplasma. 2017;**254**:685-696

[56] Babu RC, Zhang J, Blum A, Ho TH, Wu R, Nguyen HT. HVA1, a LEA gene from barley confers dehydration tolerance in transgenic rice (*Oryza sativa* L.) via cell membrane protection. Plant Science. 2004;**166**(4):855-862

[57] Ke YT, Lu CA, Wu SJ, Yeh CH. Characterization of rice group 3 LEA genes in developmental stages and under abiotic stress. Plant Molecular Biology Reporter. 2016;**34**:1003-1015

[58] Sharma A, Kumar D, Kumar S, Rampuria S, Reddy AR, Kirti PB. Ectopic expression of an atypical hydrophobic group 5 LEA protein from wild peanut, *Arachis diogoi* confers abiotic stress tolerance in tobacco. PLoS One. 2016;**11**(3):e0150609

[59] Malik MK, Slovin JP, Hwang CH, Zimmerman JL. Modified expression of a carrot small heat shock protein gene, Hsp17. 7, results in increased or decreased thermotolerance. The Plant Journal. 1999;**20**(1):89-99

[60] Katiyar-Agarwal S, Agarwal M, Grover A. Heat-tolerant basmati rice engineered by over-expression of hsp101. Plant Molecular Biology. 2003;**51**:677-686

[61] Sun W, Bernard C, Van De Cotte B, Van Montagu M, Verbruggen N.

At-HSP17. 6A, encoding a small heat-shock protein in Arabidopsis, can enhance osmotolerance upon overexpression. The Plant Journal. 2001;**27**(5):407-415

[62] Wang Y, Ying J, Kuzma M, Chalifoux M, Sample A, McArthur C, et al. Molecular tailoring of farnesylation for plant drought tolerance and yield protection. The Plant Journal. 2005;**43**(3):413-424

[63] Gharsallah C, Fakhfakh H, Grubb D, Gorsane F. Effect of salt stress on ion concentration, proline content, antioxidant enzyme activities and gene expression in tomato cultivars. AoB Plants. 2016;**8**:plw055

[64] Rahneshan Z, Nasibi F, Moghadam AA. Effects of salinity stress on some growth, physiological, biochemical parameters and nutrients in two pistachio (*Pistacia vera* L.) rootstocks. Journal of Plant Interactions. 2018;**13**:73-82

[65] Pan T, Liu M, Kreslavski VD, Zharmukhamedov SK, Nie C, Yu M, et al. Non-stomatal limitation of photosynthesis by soil salinity. Critical Reviews in Environmental Science and Technology. 2021;**51**:791-825

[66] Yue Y, Zhang M, Zhang J, Duan L, Li Z. SOS1 gene overexpression increased salt tolerance in transgenic tobacco by maintaining a higher K+/Na+ ratio. Journal of Plant Physiology. 2012;**169**(3):255-261

[67] Mian A, Oomen RJ, Isayenkov S, Sentenac H, Maathuis FJ, Véry AA. Over-expression of an Na+-and K+-permeable HKT transporter in barley improves salt tolerance. The Plant Journal. 2011;**68**:468-479

[68] Yue E, Cao H, Liu B. OsmiR535, a potential genetic editing target for drought and salinity stress tolerance in *Oryza sativa*. Plants. 2020;**9**:1337

[69] Kim S-T, Choi M, Bae S-J, Kim J-S. The functional association of ACQOS/VICTR with salt stress resistance in *Arabidopsis thaliana* was confirmed by CRISPR-mediated mutagenesis. International Journal of Molecular Sciences. 2021;**22**:11389

[70] Bhat JA, Deshmukh R, Zhao T, Patil G, Deokar A, Shinde S, et al. Harnessing high-throughput phenotyping and genotyping for enhanced drought tolerance in crop plants. Journal of Biotechnology. 2020;**324**:248-260

[71] Li X, Xu S, Fuhrmann-Aoyagi MB, Yuan S, Iwama T, Kobayashi M, et al. CRISPR/Cas9 technique for temperature, drought, and salinity stress responses. Current Issues in Molecular Biology. 2022;**44**(6):2664-2682

[72] Zhao Y, Zhang C, Liu W, Gao W, Liu C, Song G, et al. An alternative strategy for targeted gene replacement in plants using a dual-sgRNA/Cas9 design. Scientific Reports. 2016;**6**:23890

[73] Osakabe Y, Watanabe T, Sugano SS, Ueta R, Ishihara R, Shinozaki K, et al. Optimization of CRISPR/Cas9 genome editing to modify abiotic stress responses in plants. Scientific Reports. 2016;**6**:26685

[74] Park J-J, Dempewolf E, Zhang W, Wang Z-Y. RNA-guided transcriptional activation via CRISPR/dCas9 mimics overexpression phenotypes in Arabidopsis. PLoS One. 2017;**12**:e0179410

[75] Nuñez-Muñoz L, Vargas-Hernández B, Hinojosa-Moya J, Ruiz-Medrano R, Xoconostle-Cázares B. Plant drought tolerance provided

through genome editing of the trehalase gene. Plant Signaling & Behavior. 2021;**16**:1877005

[76] Shi J, Gao H, Wang H, Lafitte HR, Archibald RL, Yang M, et al. ARGOS 8 variants generated by CRISPR-Cas9 improve maize grain yield under field drought stress conditions. Plant Biotechnology Journal. 2017;**15**:207-216

[77] Ogata T, Ishizaki T, Fujita M, Fujita Y. CRISPR/Cas9-targeted mutagenesis of OsERA1 confers enhanced responses to abscisic acid and drought stress and increased primary root growth under nonstressed conditions in rice. PLoS One. 2020;**15**:e0243376

[78] He X, Luo X, Wang T, Liu S, Zhang X, Zhu L. GhHB12 negatively regulates abiotic stress tolerance in Arabidopsis and cotton. Environmental and Experimental Botany. 2020;**176**:104087

[79] Wu J, Yan G, Duan Z, Wang Z, Kang C, Guo L, et al. Roles of the *Brassica napus* DELLA protein BnaA6. RGA, in modulating drought tolerance by interacting with the ABA signaling component BnaA10. ABF2. Frontiers in Plant Science. 2020;**11**:577

[80] Zhang H, Zhang J, Wei P, Zhang B, Gou F, Feng Z, et al. The CRISPR/Cas9 system produces specific and homozygous targeted gene editing in rice in one generation. Plant Biotechnology Journal. 2014;**12**:797-807

[81] Liao S, Qin X, Luo L, Han Y, Wang X, Usman B, et al. CRISPR/Cas9-induced mutagenesis of semi-rolled leaf1, 2 confers curled leaf phenotype and drought tolerance by influencing protein expression patterns and ROS scavenging in rice (*Oryza sativa* L.). Agronomy. 2019;**9**:728

[82] Lan T, Zheng Y, Su Z, Yu S, Song H, Zheng X, et al. OsSPL10, a SBP-box gene, plays a dual role in salt tolerance and trichome formation in rice (*Oryza sativa* L.). G3: Genes, genomes. Genetics. 2019;**9**(12):4107-4114

[83] Lu Y, Tian Y, Shen R, Yao Q, Wang M, Chen M, et al. Targeted, efficient sequence insertion and replacement in rice. Nature Biotechnology. 2020;**38**(12):1402-1407

[84] Ding F, Qiang X, Jia Z, Li L, Hu J, Yin M, et al. Knockout of a novel salt responsive gene SlABIG1 enhance salinity tolerance in tomato. Environmental and Experimental Botany. 2022;**200**:104903

[85] Liu L, Zhang J, Xu J, Li Y, Guo L, Wang Z, et al. CRISPR/Cas9 targeted mutagenesis of SlLBD40, a lateral organ boundaries domain transcription factor, enhances drought tolerance in tomato. Plant Science. 2020;**301**:110683

[86] Li R, Liu C, Zhao R, Wang L, Chen L, Yu W, et al. CRISPR/Cas9-mediated SlNPR1 mutagenesis reduces tomato plant drought tolerance. BMC Plant Biology. 2019;**19**:1-3

[87] Kim D, Alptekin B, Budak H. CRISPR/Cas9 genome editing in wheat. Functional & Integrative Genomics. 2018;**18**:31-41

[88] Zheng M, Lin J, Liu X, Chu W, Li J, Gao Y, et al. Histone acetyltransferase TaHAG1 acts as a crucial regulator to strengthen salt tolerance of hexaploid wheat. Plant Physiology. 2021;**186**(4):1951-1969

[89] Liu S, Li C, Wang H, Wang S, Yang S, Liu X, et al. Mapping regulatory variants controlling gene expression in drought response and tolerance in maize. Genome Biology. 2020;**21**:1-22

[90] Zhang M, Cao Y, Wang Z, Wang ZQ, Shi J, Liang X, et al. A retrotransposon in an HKT1 family sodium transporter causes variation of leaf Na+ exclusion and salt tolerance in maize. New Phytologist. 2018;**217**(3):1161-1176

[91] Wang T, Xun H, Wang W, Ding X, Tian H, Hussain S, et al. Mutation of GmAITR genes by CRISPR/Cas9 genome editing results in enhanced salinity stress tolerance in soybean. Frontiers in Plant Science. 2021;**12**:779598

[92] Fu L, Wu D, Zhang X, Xu Y, Kuang L, Cai S, et al. Vacuolar H+-pyrophosphatase HVP10 enhances salt tolerance via promoting Na+ translocation into root vacuoles. Plant Physiology. 2022;**188**(2):1248-1263

[93] Sun T, Ma N, Wang C, Fan H, Wang M, Zhang J, et al. A golgi-localized sodium/hydrogen exchanger positively regulates salt tolerance by maintaining higher K+/Na+ ratio in soybean. Frontiers in Plant Science. 2021;**12**:638340

[94] Huang Y, Cao H, Yang L, Chen C, Shabala L, Xiong M, et al. Tissue-specific respiratory burst oxidase homolog-dependent H_2O_2 signaling to the plasma membrane H+-ATPase confers potassium uptake and salinity tolerance in Cucurbitaceae. Journal of Experimental Botany. 2019;**70**(20):5879-5893

[95] Ullah MA, Abdullah-Zawawi MR, Zainal-Abidin RA, Sukiran NL, Uddin MI, Zainal Z. A review of integrative omic approaches for understanding rice salt response mechanisms. Plants. 2022;**11**(11):1430

[96] Shailani A, Joshi R, Singla-Pareek SL, Pareek A. Stacking for future: Pyramiding genes to improve drought and salinity tolerance in rice. Physiologia Plantarum. 2021;**172**(2):1352-1362

[97] Golldack D, Lüking I, Yang O. Plant tolerance to drought and salinity: Stress regulating transcription factors and their functional significance in the cellular transcriptional network. Plant Cell Reports. 2011;**30**:1383-1391

[98] Swarup S, Cargill EJ, Crosby K, Flagel L, Kniskern J, Glenn KC. Genetic diversity is indispensable for plant breeding to improve crops. Crop Science. 2021;**61**(2):839-852

[99] Qin H, Li Y, Huang R. Advances and challenges in the breeding of salt-tolerant rice. International Journal of Molecular Sciences. 2020;**21**(21):8385

[100] Reguera M, Peleg Z, Blumwald E. Targeting metabolic pathways for genetic engineering abiotic stress-tolerance in crops. Biochimica et Biophysica Acta (BBA)-Gene Regulatory Mechanisms. 2012;**1819**(2):186-194

[101] Erdoğan İ, Cevher-Keskin B, Bilir Ö, Hong Y, Tör M. Recent developments in CRISPR/Cas9 genome-editing technology related to plant disease resistance and abiotic stress tolerance. Biology. 2023;**12**(7):1037

[102] Hafeez A, Ali B, Javed MA, Saleem A, Fatima M, Fathi A, et al. Plant breeding for harmony between sustainable agriculture, the environment, and global food security: An era of genomics-assisted breeding. Planta. 2023;**258**(5):97

[103] Fita A, Rodríguez-Burruezo A, Boscaiu M, Prohens J, Vicente O. Breeding and domesticating crops adapted to drought and salinity: A new paradigm for increasing food production. Frontiers in Plant Science. 2015;**6**:978

[104] Farooq M, Hussain M, Ul-Allah S, Siddique KH. Physiological and agronomic approaches for improving

water-use efficiency in crop plants. Agricultural Water Management. 2019;**219**:95-108

[105] Qaim M. Bt cotton in India: Field trial results and economic projections. World Development. 2003;**31**(12):2115-2127

[106] Su ZE, Zhao J, Marek TH, Liu K, Harrison MT, Xue Q. Drought tolerant maize hybrids have higher yields and lower water use under drought conditions at a regional scale. Agricultural Water Management. 2022;**274**:107978

[107] Ashraf M, Akram NA. Improving salinity tolerance of plants through conventional breeding and genetic engineering: An analytical comparison. Biotechnology Advances. 2009;**27**(6):744-752

[108] Joshi R, Sahoo KK, Singh AK, Anwar K, Pundir P, Gautam RK, et al. Enhancing trehalose biosynthesis improves yield potential in marker-free transgenic rice under drought, saline, and sodic conditions. Journal of Experimental Botany. 2020;**71**(2):653-668

[109] Niknam SR, McComb J. Salt tolerance screening of selected Australian woody species—A review. Forest Ecology and Management. 2000;**139**(1-3):1-9

[110] Kumar Y, Devi S, Phougat D, Chaurasia H, Choudhary S. Assessment of barley genotypes for salinity tolerance based on various indices under field condition. Assessment. 2024;**12**(2):586-598

[111] Maryum Z, Luqman T, Nadeem S, Khan SMUD, Wang B, Ditta A, et al. An overview of salinity stress, mechanism of salinity tolerance and strategies for its management in cotton. Frontiers in Plant Science. 2022;**13**:907937

[112] Murtaza G, Usman M, Iqbal J, Tahir MN, Elshikh MS, Alkahtani J, et al. The impact of biochar addition on morpho-physiological characteristics, yield and water use efficiency of tomato plants under drought and salinity stress. BMC Plant Biology. 2024;**24**(1):356

[113] Kumar S, Beena AS, Awana M, Singh A. Physiological, biochemical, epigenetic and molecular analyses of wheat (*Triticum aestivum*) genotypes with contrasting salt tolerance. Frontiers in Plant Science. 2017;**8**:1151

[114] Patwa N, Pandey V, Gupta OP, Yadav A, Meena MR, Ram S, et al. Unravelling wheat genotypic responses: Insights into salinity stress tolerance in relation to oxidative stress, antioxidant mechanisms, osmolyte accumulation and grain quality parameters. BMC Plant Biology. 2024;**24**(1):875

[115] Xu Y, Bu W, Xu Y, Fei H, Zhu Y, Ahmad I, et al. Effects of salt stress on physiological and agronomic traits of rice genotypes with contrasting salt tolerance. Plants. 2024;**13**(8):1157

[116] Sikder RK, Wang X, Jin D, Zhang H, Gui H, Dong Q, et al. Screening and evaluation of reliable traits of upland cotton (*Gossypium hirsutum* L.) genotypes for salt tolerance at the seedling growth stage. Journal of Cotton Research. 2020;**3**:1-13

[117] Ortiz R, Jarvis A, Fox P, Aggarwal PK, Campbell BM. Plant Genetic Engineering, Climate Change and Food Security. Copenhagen, Denmark: CCAFS Working Paper66; 2014

[118] Khan A, Pan X, Najeeb U, Tan DK, Fahad S, Zahoor R, et al. Coping with drought: Stress and adaptive mechanisms, and management through cultural and molecular alternatives in

cotton as vital constituents for plant stress resilience and fitness. Biological Research. 2018;**51**:1-17

[119] Krishnamurthy P, Vishal B, Khoo K, Rajappa S, Loh CS, Kumar PP. Expression of AoNHX1 increases salt tolerance of rice and Arabidopsis, and bHLH transcription factors regulate AtNHX1 and AtNHX6 in Arabidopsis. Plant Cell Reports. 2019;**38**:1299-1315

[120] Ji K, Kai W, Zhao B, Sun Y, Yuan B, Dai S, et al. SlNCED1 and SlCYP707A2: Key genes involved in ABA metabolism during tomato fruit ripening. Journal of Experimental Botany. 2014;**65**(18):5243-5255

[121] Chen YS, Lo SF, Sun PK, Lu CA, Ho TH, Yu SM. A late embryogenesis abundant protein HVA 1 regulated by an inducible promoter enhances root growth and abiotic stress tolerance in rice without yield penalty. Plant Biotechnology Journal. 2015;**13**(1):105-116

[122] Rasheed A, Zhao L, Raza A, Mahmood A, Xing H, Lv X, et al. Role of molecular breeding tools in enhancing the breeding of drought-resilient cotton genotypes: An updated review. Water. 2023;**15**(7):1377

[123] Nagamalla SS, Alaparthi MD, Mellacheruvu S, Gundeti R, Earrawandla JP, Sagurthi SR. Morpho-physiological and proteomic response of Bt-cotton and non-Bt cotton to drought stress. Frontiers in Plant Science. 2021;**12**:663576

[124] Kamburova V, Salakhutdinov I, Abdurakhmonov IY. Cotton Breeding in the View of Abiotic and Biotic Stresses: Challenges and Perspectives. London, UK: IntechOpen; 2022

[125] Gupta K, Karihaloo J, Khetarpal R. Biosafety Regulations for GM Crops in Asia-Pacific. Bangkok: Asia-Pacific Consortium on Agricultural Biotechnology, New Delhi and Asia-Pacific Association of Agricultural Research Institutions; 2014. pp. i-xii. 1-60

[126] Cooper M, Gho C, Leafgren R, Tang T, Messina C. Breeding drought-tolerant maize hybrids for the US corn-belt: Discovery to product. Journal of Experimental Botany. 2014;**65**(21):6191-6204

[127] Sustek-Sánchez F, Rognli OA, Rostoks N, Sõmera M, Jaškūnė K, Kovi MR, et al. Improving abiotic stress tolerance of forage grasses–prospects of using genome editing. Frontiers in Plant Science. 2023;**14**:1127532

[128] Banga SS, Kang MS. Developing climate-resilient crops. Journal of Crop Improvement. 2 Jan 2014;**28**(1):57-87

[129] Glover D. Is Bt cotton a pro-poor technology? A review and critique of the empirical record. Journal of Agrarian Change. 2010;**10**(4):482-509

[130] Balasubramani G, Raghavendra KP, Das J, Kumar R, Santosh HB, Amudha J, et al. Critical evaluation of GM cotton. In: Cotton Precision Breeding. Cham: Springer International Publishing; 2021. pp. 351-410

[131] Meissle M, Romeis J, Bigler F. Bt maize and integrated pest management-a European perspective. Pest Management Science. 2011;**67**(9):1049-1058

[132] Bonny S. Genetically modified glyphosate-tolerant soybean in the USA: Adoption factors, impacts and prospects. A review. Agronomy for Sustainable Development. 2008;**28**:21-32

[133] Paine JA, Shipton CA, Chaggar S, Howells RM, Kennedy MJ, Vernon G,

et al. Improving the nutritional value of golden rice through increased pro-vitamin a content. Nature Biotechnology. 2005;**23**(4):482-487

[134] Sadikiel MG. The adoption of genetically modified crops in Africa: The public's current perception, the regulatory obstacles, and ethical challenges. GM Crops & Food. 2024;**15**(1):185-199

[135] Lucht JM. Public acceptance of plant biotechnology and GM crops. Viruses. 2015;**7**(8):4254-4281

[136] Warwick SI, Beckie HJ, Hall LM. Gene flow, invasiveness, and ecological impact of genetically modified crops. Annals of the New York Academy of Sciences. 2009;**1168**(1):72-99

[137] Gaba Y, Pareek A, Singla-Pareek SL. Raising climate-resilient crops: Journey from the conventional breeding to new breeding approaches. Current Genomics. 2021;**22**(6):450-467

[138] Turan S, Cornish K, Kumar S. Salinity tolerance in plants: Breeding and genetic engineering. Australian Journal of Crop Science. 2012;**6**(9):1337-1348

[139] Kummari D, Palakolanu SR, Kishor PK, Bhatnagar-Mathur P, Singam P, Vadez V, et al. An update and perspectives on the use of promoters in plant genetic engineering. Journal of Biosciences. 2020;**45**:1-24

[140] Acquaah G. Conventional plant breeding principles and techniques. In: Advances in Plant Breeding Strategies: Breeding, Biotechnology and Molecular Tools. Cham: Springer; 2015. pp. 115-158

[141] Nicholl DS. An Introduction to Genetic Engineering. Cambridge: Cambridge University Press; 2023

[142] Ladics GS, Bartholomaeus A, Bregitzer P, Doerrer NG, Gray A, Holzhauser T, et al. Genetic basis and detection of unintended effects in genetically modified crop plants. Transgenic Research. 2015;**24**:587-603

[143] National Academies of Sciences, Medicine, Division on Earth, Life Studies, Committee on Genetically Engineered Crops, Past Experience, Future Prospects. Genetically Engineered Crops: Experiences and Prospects. Washington, DC: National Academies Press; 2016

[144] Eckerstorfer MF, Grabowski M, Lener M, Engelhard M, Simon S, Dolezel M, et al. Biosafety of genome editing applications in plant breeding: Considerations for a focused case-specific risk assessment in the EU. Biotech. 2021;**10**(3):10

[145] Mueller NG, Flachs A. Domestication, crop breeding, and genetic modification are fundamentally different processes: Implications for seed sovereignty and agrobiodiversity. Agriculture and Human Values. 2022;**39**(1):455-472

[146] Rosero A, Granda L, Berdugo-Cely JA, Šamajová O, Šamaj J, Cerkal R. A dual strategy of breeding for drought tolerance and introducing drought-tolerant, underutilized crops into production systems to enhance their resilience to water deficiency. Plants. 2020;**9**(10):1263

[147] Bagwan JD, Patil SJ, Mane AS, Kadam VV, Vichare S. Genetically modified crops: Food of the future. International Journal of Advanced Biotechnology and Research. 2010;**1**(1):21-30

[148] Singer SD, Laurie JD, Bilichak A, Kumar S, Singh J. Genetic variation

and unintended risk in the context of old and new breeding techniques. Critical Reviews in Plant Sciences. 2021;**40**(1):68-108

[149] Collins NC, Tardieu F, Tuberosa R. Quantitative trait loci and crop performance under abiotic stress: Where do we stand? Plant Physiology. 2008;**147**(2):469-486

[150] Munns R, Gilliham M. Salinity tolerance of crops–what is the cost? New Phytologist. 2015;**208**(3):668-673

[151] Raj SR, Nadarajah K. QTL and candidate genes: Techniques and advancement in abiotic stress resistance breeding of major cereals. International Journal of Molecular Sciences. 2022;**24**(1):6

[152] Roy SJ, Negrão S, Tester M. Salt resistant crop plants. Current Opinion in Biotechnology. 2014;**26**:115-124

[153] Zhang HX, Blumwald E. Transgenic salt-tolerant tomato plants accumulate salt in foliage but not in fruit. Nature Biotechnology. 2001;**19**(8):765-768

[154] Waltz E. Gene-edited CRISPR mushroom escapes US regulation. Nature. 2016;**532**(7599):293

[155] Tester M, Langridge P. Breeding technologies to increase crop production in a changing world. Science. 2010;**327**(5967):818-822

[156] Watson A, Ghosh S, Williams MJ, Cuddy WS, Simmonds J, Rey MD, et al. Speed breeding is a powerful tool to accelerate crop research and breeding. Nature Plants. 2018;**4**(1):23-29

[157] Hickey LT, Hafeez NA, Robinson H, Jackson SA, Leal-Bertioli SC, Tester M, et al. Breeding crops to feed 10 billion. Nature Biotechnology. 2019;**37**(7):744-754

[158] Varshney RK, Bohra A, Yu J, Graner A, Zhang Q, Sorrells ME. Designing future crops: Genomics-assisted breeding comes of age. Trends in Plant Science. 2021;**26**(6):631-649

[159] Tardieu F, Simonneau T, Muller B. The physiological basis of drought tolerance in crop plants: A scenario-dependent probabilistic approach. Annual Review of Plant Biology. 2018;**69**(1):733-759

[160] Liu W, Stewart CN. Plant synthetic biology. Trends in Plant Science. 2015;**20**(5):309-317

[161] Schaumberg KA, Antunes MS, Kassaw TK, Xu W, Zalewski CS, Medford JI, et al. Quantitative characterization of genetic parts and circuits for plant synthetic biology. Nature Methods. 2016;**13**(1):94-100

[162] Oladosu Y, Rafii MY, Samuel C, Fatai A, Magaji U, Kareem I, et al. Drought resistance in rice from conventional to molecular breeding: A review. International Journal of Molecular Sciences. 2019;**20**(14):3519

[163] Shelake RM, Kadam US, Kumar R, Pramanik D, Singh AK, Kim JY. Engineering drought and salinity tolerance traits in crops through CRISPR-mediated genome editing: Targets, tools, challenges, and perspectives. Plant Communications. 2022;**3**(6):1-27

[164] Moreira FF, Oliveira HR, Volenec JJ, Rainey KM, Brito LF. Integrating high-throughput phenotyping and statistical genomic methods to genetically improve longitudinal traits in crops. Frontiers in Plant Science. 2020;**11**:681

[165] Sun L, Lai M, Ghouri F, Nawaz MA, Ali F, Baloch FS, et al. Modern plant breeding techniques in crop improvement and genetic diversity: From molecular markers and gene editing to artificial intelligence—A critical review. Plants. 2024;**13**(19):2676

[166] Lu BR, Yang C. Gene flow from genetically modified rice to its wild relatives: Assessing potential ecological consequences. Biotechnology Advances. 2009;**27**(6):1083-1091

[167] Pilson D, Prendeville HR. Ecological effects of transgenic crops and the escape of transgenes into wild populations. Annual Review of Ecology, Evolution, and Systematics. 15 Dec 2004;**35**(1):149-174

[168] Ribichich KF, Chiozza M, Ávalos-Britez S, Cabello JV, Arce AL, Watson G, et al. Successful field performance in warm and dry environments of soybean expressing the sunflower transcription factor HB4. Journal of Experimental Botany. 2020;**71**(10):3142-3156

[169] Hill CB, Li C. Genetic improvement of heat stress tolerance in cereal crops. Agronomy. 17 May 2022;**12**(5):1-30

[170] Yu Z, Niu L, Cai Q, Wei J, Shang L, Yang X, et al. Improved salt-tolerance of transgenic soybean by stable over-expression of AhBADH gene from *Atriplex hortensis*. Plant Cell Reports. 2023;**42**(8):1291-1310

[171] Abdelsattar M, Ramadan AM, Eltayeb AE, Saleh OM, Abdel-Tawab FM, Fahmy EM, et al. Development of transgenic wheat plants withstand salt stress via the MDAR1 gene. GM Crops & Food. 2025;**16**(1):173-187

[172] Guo M, Wang XS, Guo HD, Bai SY, Khan A, Wang XM, et al. Tomato salt tolerance mechanisms and their potential applications for fighting salinity: A review. Frontiers in Plant Science. 2022;**13**:949541

[173] Sathee L, Sairam RK, Chinnusamy V, Jha SK, Singh D. Upregulation of genes encoding plastidic isoforms of antioxidant enzymes and osmolyte synthesis impart tissue tolerance to salinity stress in bread wheat. Physiology and Molecular Biology of Plants. 2022;**28**(9):1639-1655

[174] Kuzma J. Regulating gene-edited crops. Issues in Science and Technology. 2018;**35**(1):80-85

[175] Voigt B, Münichsdorfer A. Regulation of Genome Editing in Plant Biotechnology: A Comparative Analysis of Regulatory Frameworks of Selected Countries and the EU. Cham: Springer; 2019. pp. 137-238

[176] Sabat M, Tripathy A. Genetically modified and gene-edited food crops: Recent status and future prospects. In: Food Production, Diversity, and Safety under Climate Change. Cham: Springer Nature; 2024. pp. 211-222

[177] Abdul Aziz M, Brini F, Rouached H, Masmoudi K. Genetically engineered crops for sustainably enhanced food production systems. Frontiers in Plant Science. 2022;**13**:1027828

[178] Serote B, Mokgehle S, Senyolo G, du Plooy C, Hlophe-Ginindza S, Mpandeli S, et al. Exploring the barriers to the adoption of climate-smart irrigation technologies for sustainable crop productivity by smallholder farmers: Evidence from South Africa. Agriculture. 2023;**13**(2):246

[179] Zhang Y, Ding J, Wang H, Su L, Zhao C. Biochar addition alleviate the negative effects of drought and salinity

stress on soybean productivity and water use efficiency. BMC Plant Biology. 2020;**20**:1-1

[180] Chen J, Wang Y. Understanding the salinity resilience and productivity of halophytes in saline environments. Plant Science. 2024;**346**:112171

Chapter 2

Molecular Breeding Strategies for Developing Salt Stress-Tolerant Rice

Murat Aycan, Hajime Yanai, Mustafa Yildiz and Toshiaki Mitsui

Abstract

To achieve global goals for sustainable development, such as zero hunger and ensuring food security, the effect of climate change on plant production and overcoming decreasing crop yields caused by environmental stress factors must be understood. Soil salinity is a serious environmental stress factor that threatens food security, combined with global warming, reduced arable land, and the growing human population. Seven percent of the total soil areas of the world are affected by salt, which represents 33% of agricultural land. These areas are getting increased due to unplanned irrigation and aridity. It is reported that 50% of the cultivation areas will be destroyed in 25 years, depending on increased salinity. One of the promising solutions is salt stress-tolerant plant breeding to improve crop yield and quality. Currently, conventional breeding strategies are increasing attention to salt-tolerant rice development. However, it takes a long time, has a relatively low success percentage, and has primitive parental selection progress, so conventional breeding methods must be improved. Combining molecular approaches and conventional breeding methods can be a promising approach to developing salt-tolerant rice to ensure food security. In this chapter, we will explain how conventional breeding methods can be combined with modern molecular methods and new approaches can be developed.

Keywords: resilience, selection, NaCl, whole genome sequencing, SNPs, RNAseq

1. Introduction

Human activity has always been a deterministic factor in the environment. Its effects on the environment increase with knowledge of the environment [1]. Human civilization started with the development of tools and farming practices. Farming created societies, and societies brought countries until the Industrial Revolution. From the Industrial Revolution to today, humankind has gone further than before in technology, space exploration, human population, and food demand [2]. During this technological development, we have also imbalanced nature and the environment. Global climate change and environmental pollution will devastate our farming lands and practices decade by decade [3]. The increasing human population and decreasing farming areas have pushed scientists to develop high-yield new crop varieties. But in the last two decades, interests have changed to high yield and abiotic stress, such as high temperature, drought, and salinity, tolerant varieties [4]. The main problems of this century are climate change and food security. Increasing global food demand driven by rapid population growth and reducing arable farming lands requires the development of abiotic stress-tolerant crop varieties, especially salt stress-tolerant ones [5].

Salinity is a critical issue that requires careful monitoring and the development of new agricultural strategies in saline regions. Reduced plant yields and rising salinity have caused an annual economic loss of US$27.3 billion [6]. Rice is one of the salt-sensitive crops worldwide. To create novel rice genotypes with improved resistance to salt stress, it is crucial to understand the mechanism by which plants tolerate salinity and the capacity of rice to withstand it. Ensuring food security and advancing global food equality depend on this progress. Understanding the current characteristics of the farmed rice is crucial for choosing the right parental lines for breeding programs to create new high salt-tolerant genotypes [7].

To overcome these challenges, new technologies enabling the selection of individuals for parentage and the expedited selection of offspring during breeding stages are currently used in breeding programs focused on developing new salt-tolerant rice genotypes [8]. The increasing intention is to combine molecular breeding strategies with traditional breeding approaches. Current methods, such as marker-assisted selection (MAS), genome-wide association studies (GWAS), quantitative trait loci (QTL) mapping, and next-generation sequencing (NGS), have advanced abilities to identify and determine salt-tolerant genes. Genome editing tools like CRISPR/Cas9 and transcriptomic and proteomic analyses enable precise and rapid development of new salt-tolerant rice genotypes [9]. The advancement of sequencing technology serves as an indicator by identifying morphological features that correlate with differences in nucleic acid sequences within the genome. This technology's extensive application and advancement are critical factors influencing the efficacy and pace of future breeding studies. This chapter explains how integrating conventional breeding and molecular strategies offers a promising path forward in developing climate-resilient, salt-tolerant rice plants.

2. Conventional approaches to breeding for salinity tolerance

The first attempt to develop new salt-tolerant varieties starts with phenotypic selection. Genotype collections need to test salinity stress (doses depend on the crop species) tolerance capacity, and collected data can show the tolerance capacity of current genotypes. Using morphological traits such as plant height, root length, fresh weight, dry weight, and biomass; physiological characteristics such as photosynthesis and stomatal conductance; and biochemical parameters such as proline content, malondialdehyde content, superoxide dismutase activity, glutathione reductase activity, ascorbate peroxidase activity, catalase activity, and hydrogen peroxidase content can show the plants salinity tolerance level [10, 11].

After selecting natural tolerant genotypes, hybridization can be conducted. For salinity tolerance breeding, the selection of parental genotypes is critically essential. Previous studies showed that using a tolerant genotype as the maternal parent in hybridization produces more tolerant offspring than using it as the paternal parent [12]. A more critical process is the recurrent selection process after hybridization. Tolerant offspring can be screened using the same selection criteria applied to the parental lines. Morphological, physiological, and biochemical traits can be applied to selecting tolerant offspring genotypes.

Although conventional breeding is the most common approach, it faces several challenges in salinity tolerance. Often impacted by other environmental variations, salt stress is a quantitative characteristic under control by many genes. This limits the precise selection of desired features depending only on phenotype, especially for complex physiological properties such as ion balance, osmotic equilibrium, and oxidative stress control.

Furthermore, the eight to twelve years needed to produce new cultivars by traditional breeding make them poorly suited for fast-changing climatic conditions and soil salinization trends. There is minimal monitoring of salinity tolerance alleles throughout selection cycles and inadequate molecular understanding of gene-trait relationships. Both aspects contribute to reducing the effectiveness of traditional approaches.

Notwithstanding these limitations, traditional breeding is still an essential first step, particularly in molecular marker technologies. The fundamental worth of conventional techniques is shown by the successful production of many relatively salt-tolerant rice cultivars *via* pedigree selection and backcrossing.

3. Molecular tools in salinity tolerance breeding

Developing technologies and increasing the accessibility of new technologies accelerate the integration of molecular breeding techniques. Molecular breeding transformed conventional breeding by offering high-resolution tools for correctly identifying, tracking, and manipulating salt stress-related complex genes. These molecular tools enable scientists and breeders to vary genetic regions on DNA related to salt stress tolerance. Molecular tools enhance both selection and breeding efficiency.

3.1 Marker-assisted selection (MAS) and quantitative trait locus (QTL) mapping

Marker-assisted selection (MAS) is among the most useful and often utilized methods of molecular breeding. By using tightly linked molecular markers—such as SSR (simple sequence repeats), RFLP (restriction fragment length polymorphisms), and particularly SNP (single nucleotide polymorphisms)—MAS chooses for desired alleles throughout the breeding process [9]. Under stress, this method significantly reduces the requirement for thorough phenotypic screening and helps to choose breeding lines early on.

A noteworthy success story in rice is the finding and introgression of the Saltol QTL, which is found on chromosome 1 and contains the gene *OsHKT1*;5, a fundamental sodium transporter important in preserving Na^+ and K^+ homeostasis [13]. By reducing sodium buildup in shoots, *Saltol* greatly helps to increase salinity tolerance at the seedling stage. Using MAS backcrossing techniques, the *Saltol* QTL has been successfully introgressed into several high-yielding yet salt-sensitive cultivars including IR64, BPT5204, and Pusa Basmati [14]. These better lines have increased salinity tolerance and still maintain exceptional agronomic performance.

Many more quantitative trait loci (QTLs) have been discovered and confirmed in other rice populations outside of *Saltol*. As one example, found in recombinant inbred lines (RILs), qST-5.1 and qST-6.1 help in root growth, water absorption, and stress signal transduction [15]. Using pyramiding approaches for more general stress resilience, qSNC7 and qSKC1 control, respectively, the shoot Na^+ content and K^+ concentration [16]. Linkage study in controlled biparental populations has helped to find these QTLs, which are very fit for MAS in focused breeding projects. Moreover, MAS combined with background selection helps to retrieve the recurrent parent genome, hence reducing linkage drag.

3.2 Genome-wide association studies (GWAS)

Although genome-wide association studies (GWAS) provide a supplementary method by using the natural variability in large panels of rice accessions, QTL mapping has proven crucial in uncovering large-effect loci in organized populations [17]. With high-density SNP genotyping, GWAS, combined with high-density SNP genotyping, improves mapping precision and uncovers minor-effect alleles throughout the genome [18].

Recent GWAS analyses in rice have shown many salinity-related candidate genes, including *OsHAK21*, a high-affinity K^+ transporter engaged in preserving ionic balance under saline circumstances [19]. *OsNAC2*, a transcription factor that responds to stress by controlling downstream gene expression [20]. Several new transporter and regulatory genes, whose allelic variation contributes to salinity resistance, have been identified, including *OsNHX1*, a vacuolar Na^+/H^+ antiporter that controls ion compartmentalization [21].

Employing haplotype-based selection, GWAS also helps to identify haplotypes linked with desirable features, directly applicable in breeding. Furthermore, aggregating GWAS findings with transcriptome data increases candidate gene selection and functional prioritizing confidence.

3.3 Single-nucleotide polymorphism (SNP) markers and speed breeding

By providing fast, reasonably priced, highly informative markers throughout the genome, the development of SNP genotyping systems like high-density SNP arrays and genotyping-by-sequencing (GBS) has transformed rice breeding by providing rapid, cost-effective, and highly informative markers throughout the genome. In both genomic selection (GS) and marker-assisted recurrent selection (MARS) systems, these SNPs act as consistent molecular identifiers for choosing desired features.

Transfer of the *hst1* gene, a nonsense mutation in the *OsRR22* gene that results in early termination of a negative regulator, is among the most powerful uses of SNP-based MAS in salinity tolerance [22]. Initially found using MutMap, this mutation was confirmed to be a salinity tolerance-enhancing allele. Researchers rapidly produced the improved rice genotype YNU31-2-4 using SNP-based MAS in combination with biotron-assisted speed breeding, which shows lower shoot Na^+ accumulation, higher K^+ retention, increased photosynthetic efficiency, and enhanced yield performance under salt stress and elevated CO_2 conditions [22].

By use of regulated environmental conditions, speed breeding reduces generation times and enables many breeding cycles year. It greatly speeds the creation of stress-tolerant crops when paired with marker technology [23]. These molecular techniques taken together form the basis of next generation breeding projects as they provide accuracy, speed, and flexibility in handling salt stress in rice. In the next part, we investigate how genomic and transcriptomic methods facilitate the discovery of new targets for crop improvement and thereby deepen our knowledge of the stress response terrain.

4. Genomic and transcriptomic approaches

High-throughput genomic and transcriptomic technologies have given strong tools for breaking apart the intricate molecular networks underpinning salt stress reactions in rice. These methods offer a systems-level knowledge of how rice plants see, react to, and adapt to salt stress at the DNA, RNA, and gene regulating levels, hence transcending gene identification.

4.1 Whole-genome sequencing (WGS)

Discovering the whole genetic variation linked with salinity tolerance in plants has become a challenging task. Whole genome sequencing (WGS) is a potent tool to understand genetic variations related to salinity tolerance. Sequencing the whole genome of tolerant and sensitive genotypes allows researchers to identify single nucleotide polymorphisms (SNPs), insertions and deletions (indels), and structural changes perhaps underlying adaptive responses to salt stress [24]. This genome-wide differences provide the basis of association mapping, QTL discovery, and marker development for breeding programs as well as offer insightful understanding of the natural variation of rice. By providing accuracy and resolution, WGS enables the discovery of minor-effect alleles causing complex features like salinity tolerance.

Using of WGS data, comparative genomics helps scientists to compare the genomes of salt-tolerant and sensitive genotypes or mutants. Previous WGS studies provide key polymorphisms in genes including *OsHKT1*;5, which controls sodium transport and root-to-shoot ion partitioning; *OsRR22/hst1*, a response regulator whose loss-of-function mutation increases stress adaptation; and *OsNHX1*, a vacuolar Na^+/H^+ antiporter engaged in cellular ion compartmentalization have been identified by this comparison [25–27]. By use of WGS, the identification of such functional alleles has hastened the development of gene editing techniques and marker-assisted selection (MAS) pipelines, therefore arming breeders with means to directly introduce salinity tolerance characteristics into top rice cultivars.

WGS is absolutely essential tool for the *de novo* genome assembly of modern genotypes and wild rice species. Because of genetic differences from domestication and extensive breeding, these unused genetic resources can include different alleles and stress-resilient features lacking in modern cultivars. High-quality genome assemblies enable to identify new salt tolerant related genes, structural variations, and regulatory areas that could support complex features like salinity tolerance. All this new information might have influence on increasing yield and grain quality. Using pre-breeding and introgression techniques, these genomic resources may be used to expand the genetic basis of farmed rice, therefore assuring future-ready crops in the face of growing environmental difficulties.

4.2 Transcriptomic profiling (microarrays and RNA-Seq)

Recent developments in transcriptomic technolguy such as microarrays and RNA-sequencing (RNA-seq) tools have been enhanced our knowledge of the molecular reactions of the rice to salt stress. These methods assist to evaluate genome-wide gen expression patterns under various environmental condition such as salinity stress. Transcriptomic studies help to understand the fundamental regulatory mechanisms allowing rice plants to salinity tolerance by contrasting differentially express genes (DEGs) of salt-tolerant and sensitive genotypes [28]. Also, these experiments uncover and connect fundamental processes like ion transport, osmotic adjustment, oxidative stress detoxification, and ABA (abscisic acid). Also, these DEGs help to preserve cellular homeostasis during salt stress.

Transcriptome analysis has repeatedly shown many potential genes as fundamental controllers of salinity tolerance in rice. *OsSALT* helps salt exclusion from plant tissues; for example, *OsGRAS29* and *OsbHLH056* are transcription factors in control of root development and ROS homeostasis [26, 29]. Though usually connected with floral development, *OsMADSs* has also been linked with stress adaption potentially *via* cross-talk with hormone channels [30, 31]. Linked with osmotic control, *OsLPT1.16* is elevated in tolerant genotypes under salt stress [26]. Furthermore, microRNAs (miRNAs), notably miR156, is important for posttranscriptional gene control that functions downstream of *OsRR22/hst1*, thus altering stress-related gene networks and improving adaptation under salinity [26].

Transcriptomic data can provide new perspective on gene expression patterns under stress conditions by pointing out individual genes. Currently, many DEGs show tissue-specific expression—for instance, certain ion transporter genes are strongly expressed in roots whereas antioxidant genes are more active in leaves. Temporal analysis also revealed an increased response: late-response genes facilitate long-term adaptation; early-response genes involved in signaling and perception are active within hours after salt exposure [32]. Identification of important targets for gene editing, transgenic, and marker-assisted selection (MAS) approaches depends on such finely tuned expression patterns. Combining transcriptome data with proteomics, metabolomics, and epigenetics data provides a whole picture of plant stress responses and strong ideas for developing salt-tolerant rice variants with enhanced tolerance and yield [33].

5. Genome editing strategies for salinity tolerance

Recent breakthroughs in genome editing technology, transformed as a plant breeding methods by facilitating the precise and efficient alteration of genes linked to salt stress resistance. These technologies enable breeders and researchers to specifically target essential regulatory genes associated with complex characteristics like salinity tolerance, circumventing the constraints of random mutagenesis and protracted selection cycles. Among these methods, CRISPR/Cas9 (Clustered Regularly Interspaced Short Palindromic Repeats/CRISPR-associated protein 9) has emerged as the most prevalent and adaptable genome editing platform owing to its simplicity, effectiveness, and flexibility across many plant species [34].

CRISPR/Cas9 functions by directing a Cas9 nuclease to a specific DNA sequence *via* the use of a synthetic guide RNA (gRNA). This complex creates a double-stranded break at the target location, which the plant cell subsequently repairs by either nonhomologous end joining (NHEJ) or homology-directed repair (HDR) [35]. These repair mechanisms facilitate the insertion, deletion, or substitution of nucleotides, resulting in either gene knockout or accurate allele replacement. Unlike traditional breeding or chemical mutagenesis, genome editing provides a rapid, precise, and efficient method for trait development, avoiding the introduction of foreign DNA or unintended mutations in other genomic regions.

CRISPR/Cas9 was effectively used in rice to modify specific genes associated with salt stress response. Editing of *OsRR22/hst1* gene edited by CRISPR/Cas9 and yield increased in rice lines with markedly better Na^+/K^+ equilibrium and increased tolerance in salinity stress conditions [36]. Likewise, the deletion or alteration of genes such as *OsVDE*, *OsNAC3*, *OsMIR408*, and *OsMPT3* has shown enhanced ion regulation, detoxification of reactive oxygen species, and activation of stress-responsive genes [37–40]. These instances demonstrate the efficacy of CRISPR/Cas9 as a mechanism for both confirming gene function and swiftly creating climate-resilient, high-yielding rice cultivars. As genome editing advances—with the advent of base editors, prime editors, and epigenome editors—its potential to revolutionize sustainable agriculture and enhance stress-resilient crop breeding intensifies.

6. Epigenetic and regulatory mechanisms

The salinity stress sensing and response of plants are not only influenced by alterations in DNA sequence but also significantly influenced by epigenetic regulations. Epigenetics denotes heritable but reversible alterations on the genome influence gene expression without changing the nucleotide sequence. These alterations include DNA methylation, histone changes, chromatin remodeling, and the activity of noncoding RNAs. Within the framework of salt stress, these epigenetic modifications function as a dynamic and rapid response mechanism, enabling plants to recalibrate gene expression and physiological processes instantaneously, often in a tissue- and time-specific way.

DNA methylation is an epigenetic processes in rice, including the addition of methyl groups to cytosine bases, namely, in CG, CHG, and CHH contexts. This may result in either gene silence or activation, depending on the location and context of the methylation on the DNA sequence [41]. Methylation patterns can modify gene expression associated with salt stress response such as ion transport, osmotic adjustment, and reactive oxygen species detoxification-related genes under salt stress conditions. For instance, stress-induced demethylation of promoter regions may activate essential protective genes, and hypermethylation may inhibit genes that adversely control stress responses. Stress-induced methylation alterations can be transmitted between generations to generations, constituting a sort of transgenerational stress memory [42].

Besides DNA methylation, histone modifications—such as acetylation, methylation, and phosphorylation—govern chromatin accessibility that can influence gene expression patterns. Histone acetylation typically facilitates transcriptional activation, and genes involved in stress signaling, transport activities, and antioxidant defense have been correlated with elevated histone acetylation marks under salt stress conditions. Noncoding RNAs, including microRNAs (miRNAs) and long noncoding RNAs (lncRNAs), are essential posttranscriptional regulations on the salinity-responsive genes. MicroRNAs, including Osa-miR156 and Osa-miR169, regulate the stability and translation of stress-related genes under salt tress conditions [43]. Additionally, refining gene expression under salt stress conditions enhances plant adaptability and resilience. The epigenetic regulations provide interesting targets for epigenome engineering and precision breeding, facilitating the creation of rice varieties that are both salt-tolerant and capable of preserving stress memory to adapt more effectively to variable conditions.

7. Future directions

Rice breeding initiatives must vary and advance to ensure food security against climate change, soil salinization, and diminishing arable land. A key option is the increased use of wild rice species and modern genotypes, which allow to embody valuable genetic diversity. Wild rice genotypes can possess distinctive genes for salt stress tolerance that have been extinguished through domestication and traditional breeding practices. The integration of such genotypes *via* pre-breeding and introgression methods may substantially improve the adaptive potential of modern top cultivars.

Integrating the usage of artificial intelligence (AI) and machine learning (ML) into plant breeding can have significant potential to improve salinity stress tolerance in rice. Current computer technologies may expedite phenotype-genotype association research by discerning patterns and predicting markers within extensive multi-omics datasets. New AI-supported approaches may improve selection accuracy, predict genotype-by-environment interactions, and expose novel gene–trait interactions to raise the precision and efficiency of breeding for salinity tolerance. Developing novel rice crops that can resist many shocks will be essential to improve tolerance under demanding and unexpected environmental conditions.

Furthermore, new approaches based on epigenetic priming and stress memory can provide new path for developing new climate-resilient rice genotypes. These new approached include creating heritable epigenetic modifications that can allow plants to "remember" previous stress exposure and react more adeptly in future occurrences. Additionally, increasing the effectiveness of open-source bioinformatics platforms and promoting international cooperation will be essential for the rapid spreading of genetic resources, information, and tools for developing nations. Integrating technology innovation with genetic variety and worldwide accessibility, future rice breeding may achieve sustainability and resilience to increasing global food demand and environmental stresses.

8. Conclusion

Though not insurmountable, salt stress in rice farming presents a complex problem. Rice genotypes that are salt-tolerant, highly yielding, and climate-resilient may be developed by deliberately combining molecular technologies with conventional breeding and embracing omics and epigenetics. These developments help us reach the worldwide goals of sustainable agriculture, zero hunger, and climate-friendly food systems.

Particularly in salt-affected and coastal areas, salt stress continues to be a major obstacle in rice farming influencing crop production, food security, and farmer livelihoods. However, physiological and genetic complexity of salinity tolerance makes breeding challenging. More focused, effective, and environmentally friendly breeding techniques have come from advances in plant molecular biology, genomics, and biotechnology.

One of the new effective ways toward developing new salt tolerant, high-yielding rice genotypes is the strategic integration of conventional breeding with cutting-edge molecular techniques—including marker-assisted selection, genome editing, transcriptomics, proteomics, and epigenetic regulations. These modern combined approaches will not only increase stress tolerance but also help us to better understand the complex gene networks and signaling cascades engaged in salt stress adaptation.

Finally, current technological developments can integrate zero hunger, salt-tolerant rice development. Rice breeders and researchers can significantly help to solve the current problems by embracing innovation while maintaining genetic variety and ecological balance, therefore guaranteeing food security for next generations.

Acknowledgements

This work was supported by a Grant-in-Aid for JSPS Fellows (23KF0033) to MA, a Grant-in-Aid for Scientific Research (24 K08825) to TM, and a Grant for Promotion of KAAB Projects (Niigata University) from the Ministry of Education, Culture, Science, and Technology (Japan) and a Grant for Promotion of "KOME Co-creation Innovation" Project from Niigata University.

Conflict of interest

The authors declare no conflict of interest.

Author details

Murat Aycan[1*], Hajime Yanai[2], Mustafa Yildiz[3] and Toshiaki Mitsui[1]

1 Laboratory of Biochemistry, Institute for Social Innovation and Cooperation, Niigata University, Niigata, Japan

2 Japanese Sake Studies Course, Graduate School of Life and Food Science, Niigata University, Niigata, Japan

3 Department of Field Crops, Faculty of Agriculture, Ankara University, Ankara, Turkey

*Address all correspondence to: murataycan@agr.niigata-u.ac.jp

References

[1] Gallardo RK. The environmental impacts of agriculture: A review, international review of. Environmental and Resource Economics. 2024;**18**:165-235. DOI: 10.1561/101.00000166

[2] Hooke RLB, Martín-Duque JF, Pedraza J. Land transformation by humans: A review. GSA Today. 2012;**22**:4-10. DOI: 10.1130/GSAT151A.1

[3] Zou M, Niu J, Kang S, Li X, Lu H. The contribution of human agricultural activities to increasing evapotranspiration is significantly greater than climate change effect over Heihe agricultural region. Scientific Reports. 2017;**7**:8805. DOI: 10.1038/s41598-017-08952-5

[4] Shelake RM, Kadam US, Kumar R, Pramanik D, Singh AK, Kim JY. Engineering drought and salinity tolerance traits in crops through CRISPR-mediated genome editing: Targets, tools, challenges, and perspectives. Plant Communications. 2022;**3**:100417. DOI: 10.1016/j.xplc.2022.100417

[5] Kumar L, Chhogyel N, Gopalakrishnan T, Hasan MK, Jayasinghe SL, Kariyawasam CS, et al. Climate change and future of agri-food production, future foods: Global trends, opportunities, and sustainability. Challenges. 2022:49-79. DOI: 10.1016/B978-0-323-91001-9.00009-8

[6] Qadir M, Quillérou E, Nangia V, Murtaza G, Singh M, Thomas RJ, et al. Economics of salt-induced land degradation and restoration. Natural Resources Forum. 2014;**38**:282-295. DOI: 10.1111/1477-8947.12054

[7] Sackey OK, Feng N, Mohammed YZ, Dzou CF, Zheng D, Zhao L, et al. A comprehensive review on rice responses and tolerance to salt stress. Frontiers in Plant Science. 2025;**16**:1561280. DOI: 10.3389/fpls.2025.1561280

[8] Abdul Aziz M, Masmoudi K. Molecular breakthroughs in modern plant breeding techniques. Horticultural Plant Journal. 2025;**11**:15-41. DOI: 10.1016/J.HPJ.2024.01.004

[9] Aycan M, Baslam M, Mitsui T, Yildiz M. Genomics-assisted use of genetic resources for environmentally adaptive plant breeding. In: Cereal Crops. Boca Raton: CRC Press; 2023. pp. 157-174. DOI: 10.1201/9781003250845-8

[10] Aycan M, Baslam M, Mitsui T, Yildiz M. Assessing contrasting wheat (*Triticum aestivum* L.) cultivars responsiveness to salinity at the seedling stage and screening of tolerance marker traits. Journal of Plant Growth Regulation. 2024;**43**:2646-2666. DOI: 10.1007/s00344-024-11295-x

[11] Koc YE, Aycan M, Mitsui T. Exogenous proline suppresses endogenous proline and proline-production genes but improves the salinity tolerance capacity of salt-sensitive rice by stimulating antioxidant mechanisms and photosynthesis. Plant Physiology and Biochemistry. 2024;**214**:108914. DOI: 10.1016/j.plaphy.2024.108914

[12] Aycan M, Baslam M, Ozdemir B, Asiloglu R, Mitsui T, Yildiz M. Direct contribution of the maternal genotype on the transgenerational salinity tolerance in wheat (*Triticum aestivum* L.). Environmental and Experimental Botany. 2021;**192**:104648. DOI: 10.1016/j.envexpbot.2021.104648

[13] Ren Z-H, Gao J-P, Li L, Cai X, Huang W, Chao D-Y, et al. A rice quantitative trait locus for salt tolerance encodes a sodium transporter. Nature Genetics. 2005;**37**:1141-1146. DOI: 10.1038/ng1643

[14] Babu NN, Krishnan SG, Vinod KK, Krishnamurthy SL, Singh VK, Singh MP, et al. Marker aided incorporation of saltol, a major QTL associated with seedling stage salt tolerance, into oryza sativa 'pusa basmati 1121. Frontiers in Plant Science. 2017;**8**:41. DOI: 10.3389/fpls.2017.00041

[15] Kong W, Zhang C, Zhang S, Qiang Y, Zhang Y, Zhong H, et al. Uncovering the novel QTLs and candidate genes of salt tolerance in rice with linkage mapping, RTM-GWAS, and RNA-seq. Rice. 2021;**14**:93. DOI: 10.1186/s12284-021-00535-3

[16] Lin HX, Zhu MZ, Yano M, Gao JP, Liang ZW, Su WA, et al. QTLs for Na+ and K+ uptake of the shoots and roots controlling rice salt tolerance. Theoretical and Applied Genetics. 2004;**108**:253-560. DOI: 10.1007/s00122-003-1421-y

[17] Satrio RD, Fendiyanto MH, Nurrahmah N, Anofri N, Ikhsan M, Nugroho S, et al. Rice QTL hotspots related with seed grain size, shape, weight, and color based on genome wide association study and linkage mapping. Scientific Reports. 2025;**15**:21470. DOI: 10.1038/s41598-025-05814-3

[18] Chen H, Xie W, He H, Yu H, Chen W, Li J, et al. A high-density snp genotyping array for rice biology and molecular breeding. Molecular Plant. 2014;**7**:541-553. DOI: 10.1093/mp/sst135

[19] He Y, Yang B, He Y, Zhan C, Cheng Y, Zhang J, et al. A quantitative trait locus, qSE3, promotes seed germination and seedling establishment under salinity stress in rice. Plant Journal. 2019;**97**:1089-1104. DOI: 10.1111/tpj.14181

[20] Shen J, Lv B, Luo L, He J, Mao C, Xi D, et al. The NAC-type transcription factor OsNAC2 regulates ABA-dependent genes and abiotic stress tolerance in rice. Scientific Reports. 2017;**7**:40641. DOI: 10.1038/srep40641

[21] Fukuda A, Nakamura A, Tagiri A, Tanaka H, Miyao A, Hirochika H, et al. Function, intracellular localization and the importance in salt tolerance of a vacuolar Na^+/H^+ Antiporter from Rice. Plant & Cell Physiology. 2004;**45**:146-159. DOI: 10.1093/pcp/pch014

[22] Rana MM, Takamatsu T, Baslam M, Kaneko K, Itoh K, Harada N, et al. Salt tolerance improvement in rice through efficient SNP marker-assisted selection coupled with speed-breeding. International Journal of Molecular Sciences. 2019;**20**:2585. DOI: 10.3390/ijms20102585

[23] Kaur Rai N, Ravika, Yadav R, Jattan M, Karuna, Rai PS, et al. Speed breeding: A budding technique to improve crop plants for drought and salinity tolerance. In: Salinity and Drought Tolerance in Plants. Physiological Perspectives. 2023. pp. 295-313 DOI: 10.1007/978-981-99-4669-3_15

[24] Fuentes-Pardo AP, Ruzzante DE. Whole-genome sequencing approaches for conservation biology: Advantages, limitations and practical recommendations. Molecular Ecology. 2017;**26**:5369-5406. DOI: 10.1111/mec.14264

[25] Liu S, Zheng L, Xue Y, Zhang Q, Wang L, Shou H. Overexpression of OsVP1 and OsNHX1 increases tolerance to drought and salinity in Rice. Journal

Of Plant Biology. 2010;**53**:444-452. DOI: 10.1007/s12374-010-9135-6

[26] Aycan M, Nahar L, Baslam M, Mitsui T. B-type response regulator hst1 controls salinity tolerance in rice by regulating transcription factors and antioxidant mechanisms. Plant Physiology and Biochemistry. 2023;**196**:542-555. DOI: 10.1016/j.plaphy.2023.02.008

[27] Kobayashi NI, Yamaji N, Yamamoto H, Okubo K, Ueno H, Costa A, et al. OsHKT1;5 mediates Na^+ exclusion in the vasculature to protect leaf blades and reproductive tissues from salt toxicity in rice. Plant Journal. 2017;**91**:657-670. DOI: 10.1111/tpj.13595

[28] Formentin E, Sudiro C, Perin G, Riccadonna S, Barizza E, Baldoni E, et al. Transcriptome and cell physiological analyses in different rice cultivars provide new insights into adaptive and salinity stress responses, front. Plant Science. 2018;**9**:204. DOI: 10.3389/fpls.2018.00204

[29] Sahid S, Roy C, Shee D, Datta R, Paul S. Jacalin domain-containing protein OsSalT interacts with OsDREB2A and OsNAC1 to impart drought stress tolerance in planta. Environmental and Experimental Botany. 2021;**183**:104362. DOI: 10.1016/j.envexpbot.2020.104362

[30] Wu J, Yu C, Huang L, Gan Y. A rice transcription factor, OsMADS57, positively regulates high salinity tolerance in transgenic Arabidopsis thaliana and Oryza sativa plants. Physiologia Plantarum. 2021;**173**:1120-1135. DOI: 10.1111/ppl.13508

[31] Xu N, Chu Y, Chen H, Li X, Wu Q, Jin L, et al. Rice transcription factor OsMADS25 modulates root growth and confers salinity tolerance via the ABA–mediated regulatory pathway and ROS scavenging. PLoS Genetics. 2018;**14**:e1007662. DOI: 10.1371/journal.pgen.1007662

[32] Aycan M, Mitsui T. Regulation of common early and late stress responses in Rice by transcriptional and antioxidant mechanisms under salt stress. Journal of Plant Growth Regulation. 2024;**43**:4470-4489. DOI: 10.1007/s00344-024-11409-5

[33] Zhang F, Lu F, Wang Y, Zhang Z, Wang J, Zhang K, et al. Combined transcriptomic and physiological metabolomic analyses elucidate key biological pathways in the response of two sorghum genotypes to salinity stress. Frontiers in Plant Science. 2022;**13**:880373. DOI: 10.3389/fpls.2022.880373

[34] Kato-Inui T, Takahashi G, Hsu S, Miyaoka Y. Clustered regularly interspaced short palindromic repeats (CRISPR)/CRISPR-associated protein 9 with improved proof-reading enhances homology-directed repair. Nucleic Acids Research. 2018;**46**:4677-4688. DOI: 10.1093/nar/gky264

[35] Aljabali AAA, El-Tanani M, Tambuwala MM. Principles of CRISPR-Cas 9 technology: Advancements in genome editing and emerging trends in drug delivery. Journal of Drug Delivery Science and Technology. 2024;**92**:105338. DOI: 10.1016/j.jddst.2024.105338

[36] Zhang A, Liu Y, Wang F, Li T, Chen Z, Kong D, et al. Enhanced rice salinity tolerance via CRISPR/Cas9-targeted mutagenesis of the OsRR22 gene. Molecular Breeding. 2019;**39**:47. DOI: 10.1007/s11032-019-0954-y

[37] Wang X, Ren P, Ji L, Zhu B, Xie G. OsVDE, a xanthophyll cycle key enzyme, mediates abscisic acid biosynthesis and negatively regulates salinity tolerance in

rice. Planta. 2022;**255**:6. DOI: 10.1007/s00425-021-03802-1

[38] Zhang X, Long Y, Chen X, Zhang B, Xin Y, Li L, et al. A NAC transcription factor OsNAC3 positively regulates ABA response and salt tolerance in rice. BMC Plant Biology. 2021;**21**:546. DOI: 10.1186/s12870-021-03333-7

[39] Zhou J, Yuan M, Zhao Y, Quan Q, Yu D, Yang H, et al. Efficient deletion of multiple circle RNA loci by CRISPR-Cas9 reveals Os06circ02797 as a putative sponge for OsMIR408 in rice. Plant Biotechnology Journal. 2021;**19**:1240-1252. DOI: 10.1111/pbi.13544

[40] Huang S, Xin S, Xie G, Han J, Liu Z, Wang B, et al. Mutagenesis reveals that the rice OsMPT3 gene is an important osmotic regulatory factor. Crop Journal. 2020;**8**:759. DOI: 10.1016/j.cj.2020.02.001

[41] Zhou S, Li X, Liu Q, Zhao Y, Jiang W, Wu A, et al. DNA demethylases remodel DNA methylation in rice gametes and zygote and are required for reproduction. Molecular Plant. 2021;**14**:1569-1583. DOI: 10.1016/j.molp.2021.06.006

[42] Aycan M, Nahar L, Baslam M, Mitsui T. Transgenerational plasticity in salinity tolerance of rice: Unraveling non-genetic phenotypic modifications and environmental influences. Journal of Experimental Botany. 2024;**75**:5037-5053. DOI: 10.1093/jxb/erae211

[43] Wang Y, Luo Z, Zhao X, Cao H, Wang L, Liu S, et al. Superstar microRNA, miR156, involved in plant biological processes and stress response: A review. Scientia Horticulturae. 2023;**316**:112010. DOI: 10.1016/j.scienta.2023.112010

Section 2

Breeding for Biotic Stress Resistance

Chapter 3

Plant Breeding for Tolerance to Biotic Stressors Including Nematodes, Diseases, and Insect Pests

Refik Bozbuga, Pinar Arıdıcı Kara, Hatice Nilufer Yıldız, Pakize Gök Güler, Ümit Haydar Erol, Burak Özgören, Doğancan Kahya and Bekir Bülent Arpacı

Abstract

Plants are affected by a wide range of biotic stressors, including nematodes, insects, viroids, viruses, and bacteria, which are among the most damaging pathogens and pests. These causes can lead to significant yield losses, reduced fruit quality, weakened plant vigor, and, in severe cases, plant death. To control these threats, a variety of control methods have been developed. Among these, the breeding of resistant cultivars stands out as one of the most sustainable and effective strategies. Plant breeding plays a vital role in enhancing resistance to key biotic stresses and improving long-term crop productivity and resistance. This chapter presents a comprehensive overview of plant breeding approaches for tolerance to major biotic stressors, with a particular focus on nematodes, diseases, and insect pests, and discusses recent advances and challenges in developing resistant plant varieties.

Keywords: nematode, biotic stress, viroids, viruses, bacteria, plant breeding

1. Introduction

Biotic stress factors, such as nematodes, insects, viruses, bacteria, and other plant pathogens, constitute significant threats to global agricultural production. These stress factors adversely affect plant growth and development, often resulting in considerable yield losses and diminished crop quality. The presence of such stressors in numerous cropping systems emphasizes the need for integrated, sustainable, and long-term management strategies. For instance, nematodes are highly adaptable organisms inhabiting both terrestrial and aquatic ecosystems, with many species exhibiting parasitism on plants, animals, and humans [1, 2]. Likewise, a broad spectrum of pathogens, including viruses, viroids, bacteria, and insect pests, complicates the biotic stress site. Among the most sustainable and environmentally sound approaches to modifying such threats is the development of resistant cultivars.

In contrast to chemical control methods, which may have adverse environmental impacts or declining efficacy over time, breeding for resistance offers a robust and enduring solution. Resistant plants are capable of suppressing, avoiding, or tolerating pathogen and pest attacks. This chapter presents strategies aimed at enhancing resistance to major biotic stressors and explores both conventional and advanced approaches, addressing resistance against nematodes in plants, bacterial pathogens, viroids, and viral pathogens.

2. Resistance against nematodes in plants

Plant-parasitic nematodes (PPNs) affect several plants including crops. For instance, among the plant-parasitic nematodes of the order Tylenchida in apple and walnut orchards, 18 species were identified, highlighting their biodiversity and association with these fruit tree hosts [3]. Plants such as wheat, pistachio, barley, grapevine, tomato, and cotton, are affected by several common species and six distinct nematode species, thereby nematode diversity impacting major agricultural crops [4]. Plant-parasitic nematodes (PPNs) affect a wide variety of economically valuable crops, including members of the Solanaceae, Fabaceae, Malvaceae, Amaranthaceae, and Poaceae families and these nematodes often exhibit high virulence and can infect diverse hosts by evading or suppressing plant immune responses through complex infection strategies [5]. PPNs are multicellular organisms that damage plants by using a stylet to feed on cells, acting as either ectoparasites or endoparasites, and causing symptoms such as stunted growth, wilting, and increased susceptibility to other pathogens [6]. PPNs including cyst and root-knot nematodes, are major agricultural pests that induce specialized feeding cells in plant roots, while plants may control to plant-parasitic nematodes invasion through multi-layered immune responses, including pattern-triggered immunity *via* recognition of pathogen- and damage-associated molecular patterns by pattern recognition receptors, and effector-triggered immunity *via* intracellular leucine-rich repeat Nucleotide-binding leucine-rich repeat (NLR) receptors; these responses involve anti-nematode enzyme secretion, production of toxic compounds, cell wall strengthening, reactive oxygen species generation, nitric oxide production, and hypersensitive cell death, collectively aiming to restrict nematode development and protect plant health [7]. Some of the most economically significant plant-parasitic nematodes, root-knot nematodes (RKNs) and cyst nematodes (CNs), both reprogrammed host root cells to create specialized feeding structures essential for their development and reproduction; however, while root-knot nematodes (RKN) juveniles invade near the root tip and migrate intracellularly to the vascular cylinder where they induce the formation of multinucleate giant cells *via* repeated nuclear division without cytokinesis cyst nematodes CNs employ a distinctly different strategy to establish their feeding sites, reflecting the divergent molecular and cellular mechanisms these nematodes use to exploit host plant systems [8, 9]. Sedentary endoparasitic nematodes, such as cyst (*Heterodera* and *Globodera* spp.) and root-knot (*Meloidogyne* spp.) nematodes, damage crops by inducing specialized feeding cells—syncytia or giant cells within roots that serve as nutrient sources; these nematodes secrete effectors from esophageal glands to manipulate host proteins, reprogramming cell development and immunity, with recent research identifying plant targets to improve crop resistance and deepen understanding of root development [10]. Microscopic sedentary nematodes establish a biotrophic relationship with plants by inducing root cells to form enlarged, multinucleate feeding cells through effectors delivered *via* a stylet; these effectors manipulate host cell development, physiology, and defense by targeting nuclear processes, with recent genomic and functional studies revealing their roles in reprogramming host cells to support nematode feeding and survival [11]. Several economically significant nematode genera, including root-knot, root lesion, and cyst nematodes from the Heteroderidae family, damage crop plants, making effective management strategies essential; genetic engineering has emerged as a valuable approach for enhancing nematode resistance in crops by incorporating natural resistance genes, proteinase inhibitors, anti-nematode proteins, RNA interference targeting nematode effectors, and manipulating gene expression to develop transgenic plants with improved protection against PPNs [12]. As plant-parasitic nematodes (PPNs) are required to break the cell wall to access nutrients, the strengthening of the cell wall has been identified as an effective defensive strategy, serving as a physical barrier that inhibits nematode entry and feeding [7].

When nematodes meet plant tissue, the plant's first line of defense involves the recognition of damage-associated molecular patterns (DAMPs) and pathogen-associated molecular patterns (PAMPs) by cell surface-localized pattern recognition receptors (PRRs), which detect signals such as bacterial flagellin or fungal chitin and activate pattern-triggered immunity (PTI), a crucial early immune response against microbial invasion [13, 14]. Several damaging nematode species induce root cell redifferentiation to create specialized feeding structures, and while nematicides dominate current control in intensive farming, alternative genetic strategies are essential; natural resistance genes are used in breeding programmes, with ongoing research into their molecular mechanisms, while transgenic methods, including proteinase inhibitor expression and plant-delivered RNA interference targeting essential nematode genes, offer promising new avenues for effective nematode resistance [15]. Nucleotide-binding leucine-rich repeat (NLR) proteins are also essential in detecting plant-parasitic nematodes, with the majority of NLRs responsible for this recognition being encoded by specific resistance (R) genes, thereby contributing significantly to the plant's immune defense against nematode invasion [16]. Numerous nematode resistance genes cloned in plants often resemble other resistance genes and typically trigger host cell death at or near the nematode feeding site, with the timing and location of this response depending on the specific gene and nematode involved; although genetic studies suggest single nematode genes may influence plant resistance, the corresponding nematode effectors remain unidentified, yet advances in plant and nematode genetics, genomics, and RNA silencing techniques are enhancing our understanding of the molecular mechanisms governing these complex plant–nematode resistance interactions [17]. Isolated from plants from several nematode resistance genes, all providing protection against sedentary endoparasites, with the first cloned gene being Hs1pro-1 from sugar beet, which confers resistance to the sugar beet cyst nematode and encodes a protein unlike typical plant genes; however, other genes such as Gro1-4, Mi-1, Gpa2, and Hero A cloned from tomato and potato relatives, belong to the nucleotide-binding site-leucine-rich repeat (NBS-LRR) class of resistance genes, with Mi-1 and Hero A offering broad-spectrum resistance to multiple root-knot and potato cyst nematode species, while Gpa2 and Gro1-4 provide more specific resistance, differing in their N-terminal structures, where Mi-1, Hero A lack a Toll-interleukin receptor (TIR) domain, but Gro1-4 contains one, highlighting structural diversity within this gene family [17]. The Mi gene, encoding an NBS-LRR resistance protein in many tomato cultivars, provides resistance to root-knot nematodes and potato aphids but is temperature-sensitive, prompting the discovery of additional distinct resistance genes in *Lycopersicon peruvianum* to broaden nematode resistance in crops [18]. The identification of leucine-rich repeat (NLR)- and pattern recognition receptors (PRR)-type receptors has advanced understanding of plant immune responses to plant-parasitic nematodes (PPNs), yet many PPN-derived pathogen-associated molecular patterns (PAMPs), effectors, and their receptors remain unknown, leaving key gaps in linking nematode recognition to specific immune activation, and underscoring the urgent need for molecular research to improve sustainable control strategies for safeguarding global food production [7]. Genetic resistance is used against nematodes in some crops, but fewer resistant cultivars succeed due to limited genetic variation in hosts and nematodes; single dominant resistance genes often lack durability, especially given domestication's impact on genetic diversity, so understanding complex plant–nematode gene interactions and co-evolution is crucial for effective resistance gene identification and breeding [19].

3. Resistance and breeding strategies against bacterial pathogens

Bacterial diseases caused severe crop losses [20]. There are a few chemicals that can be effective in controlling plant bacterial diseases [21]. The best antibacterial compounds, antibiotics, have very limited use in plant agriculture [22]. Copper compounds and the antibiotic streptomycin are effective chemicals in controlling bacterial diseases [22]. Although these chemicals are effective in controlling bacterial diseases, their applications in high numbers caused resistance in bacterial populations [23]. Copper resistance occurs by horizontal gene transfer of copper resistance genes through conjugation of copper resistance plasmids; this phenomenon has been widely observed in *Xanthomonas* species [24]. Another example is streptomycin resistance in *E. amylovora*, which occurred by chromosomal mutations that altered the ribosomal protein target of the antibiotic [25]. These genes encode aminoglycoside phosphotransferase enzymes that convert streptomycin to a non-toxic form [26]. Breeding is serving effective, environmentally safety and economical way of control compared to chemical control [27]. Host plant resistance can be defined as a mechanism to reduce disease severity [28]. Breeding studies against plant bacterial diseases are very important to prevent yield losses in agricultural production and to support sustainable agriculture [29]. Disease management in breeding is achieved by targeting pathogen virulence or genetic manipulation of host plants [30]. In general, the plant's immune response to biotic stresses first begins with the recognition of the pathogen by the plant recognize PAMPs (Pathogen-Associated Molecular Patterns), which are surface proteins of bacteria [31]. When plant cells perceive bacterial signals such as flg22 (part of flagellin), they initiate a series of defense, such as cell wall thickening, production of antimicrobial compounds or oxidative burst [32].

Effector proteins secreted by some bacteria into the plant cell (e.g. *via* the T3SS) R (resistance) genes present in the plant recognize these effectors [33]. This recognition usually results in a hypersensitive response (HR) [34]. Controlled death (necrosis) of infected cells prevents the spread of the bacteria, called as Effector-Triggered Immunity [35]. Two plant phenolic compounds, o-coumaric acid (OCA) and t-cinnamic acid (TCA), have been identified as inducers of the T3SS genes of *Dickeya dadantii* [36]. Resistance genes that recognize effectors specific to bacteria are in the NBS-LRR (nucleotide-binding and leucine-rich repeats) structure [37]. Integration of R-gene groups into new cultivars has been a breeding strategy, for example, Xa3, Xa4, and Xa21 genes in the rice-bacterial blight (*Xanthomonas oryzae* pv. *oryzae*) system [38]. When a plant is exposed to a pathogen, it develops a defensive memory and responds more strongly to future attacks [39]. This spreads to other tissues with a systemic signal (systemic acquired resistance—SAR) and many phytohormones now regulate induced defense signalling, salicylic acid (SA) and jasmonic acid (JA), which are phytohormones that regulate various developmental processes in plants, such as flowering, root growth, flower nectar secretion, senescence, development, cell growth, trichome development, and thermogenesis, are the best known phytohormones for activating plant defense signalling [40]. Human-Handed Intervention (Breeding and Genetic Engineering) Selection of naturally resistant individuals, included in breeding programmes [41]. Defense genes can be edited, or genes that increase sensitivity can be silenced with techniques such as CRISPR/Cas9 [42]. Bacterial disease resistance strategies based on CRISPR/Cas9 targeted modification of susceptibility genes in important crops [43]. In the study where lines with deletions in the effector binding elements (EBE) of AvrXa7 were used, it was reported that they showed resistance against the Xoo strain [44]. Thus, CRISPR-Cas9 was shown to be usable against common *Xanthomonas oryzae* pv. *oryzae* strains and bacterial blight [45]. Transcription activator-like effector (TALE) proteins secreted by *Xanthomonas* spp. and *R. solanacearum* are transported into the plant nucleus, bind to effector binding elements (EBE) DNA fragments, and with the help of transcription accessory proteins activate the transcription of susceptibility (S) genes for disease development [46]. Some plants have evolved mechanisms to avoid the activation of S genes by modifying the EBE promoter sequences [47]. The most common bacterial diseases caused by *Xanthomonas* spp., *Pseudomonas syringae*, *Ralstonia solanacearum,* and *Erwinia amylovora* are studied in breeding studies [48, 49]. Breeding and genetic programme studies were conducted to develop *Phaseolus vulgaris* bean cultivars tolerant to the bacterial pathogens *Pseudomonas phaseolicola*, *Xanthomonas phaseoli* and *Corynebacterium flaccumfaciens* [50]. On Eggplant (*Solanum melongena* L.), two hybrids, 'VNR-218' × 'BCB-11' and 'Arka Nidhi' × 'KS-331', having large vascular area, were selected due to their high yielding and low percentile disease index (PDI) values for bacterial wilt disease [51]. Larger vascular area in the vascular bundle is considered in hybrid selection for resistance to bacterial wilt disease [51]. Wild *Malus* genotypes are found to be stronger than domestic ones against *Erwinia amylovora* by replacing a cysteine with a serine in the effector protein sequence of AvrRpt2EA, which is the effector protein in pathogenesis [52]. Like other approaches, there are significant limitations to the development and commercial application of genetic resistance [53]. Bacterial disease usually evolves very quickly, in which case it can lead to rupture [54]. Environmental factors may affect the durability of breeding crop varieties [55]. Screening plant germplasm for resistance and incorporating it into commercial varieties is a lengthy and challenging process, further complicated by reduced biodiversity due to habitat loss, sexual incompatibility, and complex genetic structures [56].

4. Resistance and breeding strategies against viroids

Viroids are small, single-stranded RNA (ssRNA) molecules that are covalently closed, circular in structure, highly structured, and range in size from approximately 250 to 400 nucleotides, and they lack protein-coding capacity [57, 58]. Despite these characteristics, viroids are capable of replicating independently in susceptible host plants without the need for any helper virus and can be systemically transmitted throughout the plant [58, 59]. They use host proteins for the replication, trafficking, and processing of their replication intermediates [57, 58], a feature that differentiates them from plant viruses [60]. The exact mechanisms by which these non-coding RNA agents induce symptoms in plants remain incompletely understood [61].

During viroid infection, the host plant's RNA silencing machinery generates viroid-derived small RNAs (vd-sRNAs) of 21–24 nucleotides, which are incorporated into AGO proteins and these vd-sRNAs are believed to mediate the degradation of viroid RNAs, thus acting as part of a plant antiviral defense response [61, 62]. However, these small RNAs may also cleave host mRNAs that encode essential proteins for plant development or cellular homeostasis, potentially leading to observable symptoms [61]. Furthermore, viroids are thought to interfere with innate immunity, hormonal signaling, and protein translation mechanisms in host plants and may also trigger epigenetic changes in host DNA [58]. Although certain plant groups are non-hosts to viroids, no stable or durable resistance has been identified to date among known viroid host species [63].

Pathogenicity, Evolution, and Resistance of Viroids in Plants: Viroids are the smallest known infectious agents, ranging from 250 to 400 nucleotides in size, following their discovery, it is reported that they possess the potential to cause serious diseases in a wide range of monocotyledonous and dicotyledonous, herbaceous and woody plant species, including numerous economically important crop and ornamental plants [60, 64]. Viroids characterized to date fall within two families: Pospiviroidae, which includes Potato spindle tuber viroid (PSTVd) and is defined by the presence of a central conserved region (CCR). Replication in Pospiviroidae occurs in the nucleus *via* host DNA-dependent RNA polymerase II, producing double-stranded RNA intermediates that accumulate during replication [57]. The infection cycle of all viroids includes cellular entry, replication, intercellular and systemic movement, and dissemination throughout all plant organs *via* the vascular system, ultimately leading to symptom development as the viroid load increases [65].

Since their initial identification, viroid RNAs have been thought to act as abnormal regulatory RNAs [66, 67]. Although significant progress has been made in understanding their biochemical properties, many aspects of viroid pathogenicity remain unclear. The origin of viroids is also uncertain and it is proposed that viroids may have originated from host cellular RNAs, plasmids, or introns and could be remnants of pre-cellular RNA evolution [68–70]. Due to the presence of ribozymes and their non-coding nature, members of the Avsunviroidae family are sometimes considered "living fossils" of the ancient RNA world [68, 71]. Interestingly, viroids are known to infect only members of the plant kingdom and some studies have noted partial structural similarities between viroids and the hepatitis delta virus (HDV), a satellite virus of hepatitis B virus (HBV). However, HDV RNA is larger and requires the expression of viral antigens critical to its replication cycle [67, 72, 73]. Viroid pathogenicity is a complex phenomenon influenced by both the viroid and host genomes, and the presence of different viroid variants within the same host can result in latent (asymptomatic) infections [74] or in a range of symptoms from mild to severe [75]. The macroscopic symptoms of viroid infections are often similar to those caused by plant viruses and may include stunting, epinasty, vascular discolouration and clearing, leaf distortion and mottling, chlorotic or necrotic spots, cankers, bark scaling and cracking, and deformations of tubers, flowers, and fruits, in rare cases, infections can result in plant death [75]. The possibility that viroids have existed as plant pathogens for centuries has led researchers to speculate that they may have been responsible for historical disease outbreaks [60, 68]. The recovery of viroid-associated diseases in the twentieth century may be attributed to the intensification of horticultural practices, widespread use of susceptible genotypes, and insufficient diagnostic methods—factors that highlight the need for improvements in both agricultural management and diagnostic capabilities [76, 77].

Given the broad host range of viroids and their ability to persist in both symptomatic and asymptomatic forms [76, 77], incorporating molecular techniques and transgenic approaches alongside traditional breeding methods may improve disease management outcomes, to date, potato spindle tuber viroid (PSTVd) and Chrysanthemum stunt viroid (CSVd), both of the Pospiviroidae family, are among the most extensively studied viroid pathogens [67]. In chrysanthemum cultivars with low CSVd titres, selfing experiments reveal that some progeny displayed strong resistance to CSVd [67]. This resistance is also expressed in certain F1 hybrids generated by crossing the resistant cultivar with two susceptible ones [78]. Furthermore, it is observed that CSVd do not replicate in the shoot apical meristems or leaf primordia of resistant plants, suggesting the presence of additional resistance mechanisms [79].

Non-transgenic methods employed for viroid control include several strategies that have been used to manage plant viroid pathogens to date. Application of 5–8°C low-temperature conditions and low light treatments followed by meristem culture has resulted in 54% healthy plant regeneration, with 29% of meristems producing viroid-free plants in the case of PSTVd [80]. Other approaches include the *in vitro* culture of shoot tips to obtain viroid-free mother plants, as well as *in vitro* shoot-tip grafting to improve sanitization efficiency and, additionally, cross-protection strategies, whereby a mild strain is used to elicit resistance *via* host defense responses (e.g. protein-mediated or RNA-mediated resistance), and methods such as electrotherapy may be used either individually or in combination [67, 81]. Furthermore, the application of salicylic acid (SA) and jasmonic acid (JA), plant hormones known to play vital roles in plant defense against biotic and abiotic stressors [82], has shown promise. For example, in *Gynura aurantiaca* plants infected with CEVd, external SA treatment is observed to significantly delay disease progression and alleviate symptoms, correlating with increased endogenous SA levels [83]. Similarly, SA application improves tolerance to PSTVd in the Rutgers tomato cultivar [71, 84].

Transgenic strategies for conferring viroid resistance in plants seek to employ natural defense mechanisms including the production of resistance proteins (R-gene products), phytoalexins, reactive oxygen species (ROS) involved in hypersensitive responses, and RNA silencing pathways [67]. These intrinsic mechanisms are often insufficient for resisting pathogenic microorganisms, and while viruses typically encode multifunctional RNA silencing suppressors to overcome plant defenses, viroids differ in that they do not encode any proteins, and thus cannot function as silencing suppressors themselves [85]. The first plant protection study against viroid infection employed an antisense RNA strategy that introduced a hammerhead ribozyme targeting the negative-strand RNA of PSTVd into potato plants [86, 87]. Of the 34 transgenic lines, 23 (approximately 68%) displayed high levels of resistance, with no detectable PSTVd RNA, and resistance was also retained in vegetative tubers [81]. Three transgenic tomato lines expressing a hairpin RNA (hpRNA) structure based on PSTVd sequences, two of which showed resistance to PSTVd infection. Although viroid RNAs in infected plants were not completely silenced, the derived hpRNA-triggered siRNAs (hp-siRNAs) effectively targeted PSTVd RNA [88]. Pac1 protein efficiently cleaves long dsRNAs into short oligonucleotides, serving as substrates for small hpRNAs. Transgenic expression of Pac1 protein conferred resistance to several ssRNA viruses [89]. Similarly, five transgenic potato lines expressing Pac1 and inoculated with PSTVd exhibited resistance, and viroid transmission *via* seed tubers is reduced, and all transgenic plants reportedly maintained healthy growth without abnormal phenotypes, indicating no adverse physiological effects from Pac1 overexpression [90]. Genome editing technologies are increasingly considered valuable tools in producing viroid-tolerant or -resistant crops that particularly, SDN-1-type genome edits, small modifications indistinguishable from those occurring naturally or *via* conventional breeding, and already present in some commercial crops, are regarded as practical and further research and development in this area are strongly recommended [71].

Since the initial identification of viroids, various strategies have been explored to prevent their transmission from infected to healthy plants [91]. Measures to mitigate the damage caused by viroid infection include screening for naturally resistant cultivars and breeding for resistance, using mild strains to induce resistance and prevent superinfection, and developing transgenic plants through advanced tools such as antisense and sense RNAs, hammerhead ribozymes, and RNA interference and these approaches have yielded varying degrees of success and resistance levels [90].

However, the secondary structures of viroid RNAs, their interactions with host proteins, and their inherent stability can enable viroids to evade host RNA silencing mechanisms, thereby limiting the effectiveness of RNA-based control methods [67]. Moreover, there is no universal strategy that can be implemented to protect all host plants from every viroid species [67, 92].

5. Resistance and breeding strategies against viral pathogens

Viruses, as one of the most destructive plant pathogens causing severe yield losses and sometimes total crop failure, depend entirely on host cellular machinery for replication and movement, rendering chemical control largely ineffective, and thus underscore the critical need for developing virus-resistant cultivars as the most reliable and sustainable approach, given the limited efficacy and long-term viability of traditional control strategies. Among plant resistance mechanisms, extreme resistance (ER), a symptomless, near-complete defense that rapidly restricts virus replication and movement at the infection site without triggering cell death, relies on nucleotide-binding leucine-rich repeat (NLR) immune receptors such as the TIR-domain NLR Rysto in potato, which confers durable ER to Potato virus Y (PVY) *via* EDS1 and NRG1 signalling even under elevated temperatures [93], and the Rsv1 gene in soybean, which mediates ER to Soybean mosaic virus (SMV) through a complex network involving GmEDR1, GmEDS1, GmHSP90, GmPAD4, and WRKY transcription factors as identified through virus-induced gene silencing (VIGS) [94]. The Rysto gene, originally introgressed from *Solanum stoloniferum* into cultivated potato, has been widely used in European breeding programmes as it is mapped Rysto to chromosome XII using CAPS and SSR markers, showing its monogenic, dominant inheritance and utility across multiple germplasm sources [95]. It is dissected the inheritance of extreme resistance (ER) in the potato cultivar 'Barbara', revealing two independent loci: Rysto conferring ER to both PVY and PVA, and a separate Ra gene conferring ER only to PVA; Segregation analyses in F1 and BC1 populations confirmed their independent actions and additive resistance profiles [96]. In addition, in the pepper plant, the Pvr4 gene from *Capsicum annuum* cv. CM334 confers broad-spectrum extreme resistance (ER) to potyviruses such as PVY and PMMoV, with its expression varying between hypersensitive response (HR) and ER depending on the viral isolate and tissue type, and notably, despite widespread use, virulent isolates remain rare under natural conditions [97]. ER and HR-like responses in potato to

Tobacco rattle virus (TRV) are mediated by a single viral protein—the movement protein. Using a DsRed-tagged TRV clone, they confirmed that ER acts at the single-cell level, suppressing replication without triggering HR [98]. Alfalfa mosaic virus (AIMV) is controlled by a recessive gene, with resistance being temperature sensitive; Although plants showed systemic resistance at moderate temperatures, virus spread occurred at higher temperatures that genetic analysis suggested a monogenic, recessive control [99]. The hypersensitive response (HR) is a well-characterized form of programmed cell death (PCD) in plants that rapidly confines viral infection through localized necrotic lesions, restricting replication and spread, and although it shares features with animal apoptosis—such as DNA fragmentation and ion fluxes—it operates through distinct plant-specific pathways, indicating the existence of multiple unique cell death mechanisms in plants [35]. HR is often triggered through gene-for-gene interactions, in which host resistance (R) genes recognize specific avirulence (Avr) determinants encoded by pathogens; in viral systems, several viral proteins act as HR elicitors—for instance, in Turnip mosaic virus (TuMV), the P3 protein alone induces HR-like necrosis in *Arabidopsis thaliana* ecotype *Landsberg erecta,* which carries the corresponding dominant resistance gene [100]. Similarly, in tomato spotted wilt virus (TSWV)-resistant *Capsicum chinense*, the N protein has been identified as the HR elicitor, inducing chlorotic and necrotic lesions accompanied by DNA fragmentation, a hallmark of PCD [101]. Three amino acid changes within the NIa protease domain were sufficient to overcome plum pox virus (PPV) resistance mediated by HR in resistant plum cultivars, underscoring the importance of proteolytic processing in viral evasion of plant defense [102]. Some viruses have evolved mechanisms to suppress HR, for example, the transcriptional activator protein (TrAP) of *Tomato leaf curl New Delhi virus* (ToLCNDV) was shown to counteract HR induced by its nuclear shuttle protein (NSP), with the zinc finger domain and nuclear localization signal of TrAP playing crucial roles in this suppression [103]. WRKY proteins are central regulators of transcriptional reprogramming during pathogen challenge, and in *Capsicum annuum*, the WRKY IIa group transcription factor CaWRKYd has been shown to regulate HR in response to Tobacco mosaic virus (TMV) infection, as its silencing reduces the expression of HR- and defense-related genes and suppresses HR-associated cell death [104]. Auxin signalling has emerged as a contributor to HR-mediated resistance. In *Capsicum chinense* infected with tomato zonate spot virus (TZSV), transcriptome and VIGS analyses identified genes involved in auxin-induced signalling that were upregulated during HR, suggesting hormonal integration in viral defense responses [105]. Temperature has a significant influence on the effectiveness of HR; the Tsw-mediated HR to TSWV in pepper plants was found to be temperature-sensitive, with resistance being compromised at 32°C. This thermosensitivity is especially pronounced in young plants and in heterozygous backgrounds, indicating the importance of both genotype and developmental stage [106]. The Arabidopsis ecotype C24 demonstrates HR to *Cucumber mosaic virus* strain Y (CMV-Y) but not to strain O (CMV-O) that this resistance is mediated by a single dominant gene influenced by a modifier gene and is shown to depend on the RNA3 component of CMV, highlighting the specificity of the HR to viral genome segments [107]. Field-level studies also report durable HR-based resistance in commercial cultivars that most resistant plum trees carrying HR to PPV maintain resistance under field conditions, occasional isolates can overcome this response, warranting continued surveillance and genetic diversification of resistance sources [102]. Pepper (*Capsicum* spp.), a globally cultivated crop threatened by numerous plant viruses such as CMV, TSWV, PVY, and ToBRFV, has traditionally relied on classical resistance genes for protection, though their durability is often undermined by virus evolution, environmental stress, and vector pressure, prompting interest in tolerance, an alternative strategy whereby plants sustain acceptable growth and yield despite systemic infection, not by preventing viral accumulation but by mitigating the physiological damage it causes [108]. The Indian *Capsicum annuum* accession 'Perennial' has been identified as a valuable donor of *CMV* tolerance, which is quantitatively inherited and incompletely dominant, as breeding lines exhibit variable symptom suppression upon mechanical inoculation, and ELISA assays reveal no significant correlation between viral load in upper leaves and symptom expression, suggesting that tolerance may involve a recovery mechanism rather than direct suppression of viral replication [109]. Recent findings reveal that *Capsicum annuum* genotypes carrying L resistance alleles (L1, L3, and L4) mount a hypersensitive-like response to ToBRFV infection that limits systemic viral spread, while certain cultivars lacking classical L alleles still exhibit tolerance by allowing systemic infection without visible fruit symptoms, and even in susceptible lines, infection of reproductive tissues does not necessarily lead to symptomatic fruit, indicating a disconnect between viral presence and symptom expression [110]. In *Capsicum baccatum*, the novel accession PIM26-1 exhibited high tolerance to both wild-type and resistance-breaking (TBR) isolates of TSWV, with qRT-PCR-based viral quantification, Kaplan–Meier infection timelines, and symptom progression analyses revealing delayed infection and minimal symptom development—none of the plants showed severe symptoms to either isolate—suggesting the presence of tolerance mechanisms distinct from classical resistance genes such as Tsw, and offering a promising avenue for breeding durable TSWV control [111]. Introgression from tolerant donor lines, such as Perennial [109] and PIM26-1 [111], into elite cultivars *via* pedigree and backcrossing remains a practical method [112]. The identification of genomic regions linked to tolerance enables marker-assisted and genomic selection, while plant viruses counter with viral suppressors of RNA silencing (VSRs), yet RNA interference (RNAi)—a potent, sequence-specific antiviral mechanism involving Dicer-like enzymes processing viral dsRNA into siRNAs that guide Argonaute proteins to silence viral genomes post-transcriptionally or transcriptionally—remains highly effective through transgenic dsRNA expression and gene editing of host susceptibility factors like eukaryotic translation initiation factors (eIFs) [112]. An effective RNAi-based approach for durable, broad-spectrum resistance involves silencing host susceptibility genes essential for viral replication, as demonstrated in melon, where RNAi-mediated knockdown of the Cm-eIF4E gene selectively reduced its mRNA and produced siRNAs without affecting the Cm-eIF(iso)4E isoform, resulting in transgenic lines resistant to four key potyviruses: CVYV, MNSV, MWMV, and ZYMV, confirming host-directed gene silencing of translation initiation factors as a powerful strategy for multi-virus resistance [113]. A robust RNAi strategy involving transgenic expression of viral gene-derived inverted repeats forming hairpin RNAs processed into virus-specific siRNAs was demonstrated in potato, where a synthetic 600-bp hairpin construct of fused coat protein sequences driven by the 35S promoter conferred nearly 100% resistance to Potato virus X, Potato virus Y, and Potato virus S over two crop seasons, highlighting RNAi's multiplexing potential for stacked virus resistance in crops [114]. The CRISPR/Cas9 system is composed of a Cas9 endonuclease guided by a synthetic single guide RNA (sgRNA) that directs cleavage at specific DNA sequences marked by a protospacer adjacent motif (PAM), typically "NGG", this RNA-guided system allows precise double-stranded breaks in both host and viral genomes, leading to gene knockout, disruption, or mutation [115]. A key application of CRISPR/Cas9 involves directly targeting viral DNA genomes, as shown in tomato and *Nicotiana benthamiana* where stable expression of Cas9 and sgRNAs against Tomato yellow leaf curl virus coat protein and replicase genes markedly reduced viral load and symptoms, with resistance inherited across generations [116]. Similarly, effective interference against Cauliflower mosaic virus was demonstrated through multiplexed sgRNAs targeting the viral coat protein gene, producing edited viral genomes with frame-shift mutations that rendered the coat proteins non-functional [117]. Recessive resistance to RNA viruses, particularly potyviruses, often involves the eukaryotic translation initiation factor eIF4E gene family, and CRISPR/Cas9 editing of these host factors enables broad-spectrum resistance without targeting the viral genome, reducing resistance breakdown risk, as reviewed in potato for Potato virus Y, Potato virus X, and Potato leaf roll virus [118]. eIF4E1 knockout in the tetraploid cultivar Desirée expanded resistance to *PVYNTN*, a strain not covered by the dominant Ny gene, with KO lines showing reduced viral load and milder symptoms, demonstrating the viability of pyramiding recessive resistance alleles *via* genome editing [119]. CRISPR/Cas systems also offer the potential for simultaneous targeting of multiple viral or host genes, demonstrating multiplexing capability as shown by the use of RNAi constructs to target multiple potato viruses [114]. CRISPR/Cas technology enhances multi-site genome editing with greater specificity, while DNA-free delivery methods like protoplast transfection with ribonucleoprotein complexes are being explored to address regulatory barriers and public concerns about genetically modified organisms [115, 118].

6. Resistance mechanisms against insects in plants

Protecting cultivated plants from diseases and pests throughout the growing season is an essential factor during the vegetation process [120]. Major challenges in agriculture include the reduction of arable land and the inability to obtain high-quality yields per unit area, resulting in significant yield losses. A well-designed plant health programme plays a crucial role in minimizing these losses [121]. Despite the implementation of various control methods, it has been reported that approximately 48% of products are lost due to insects, pathogens, weeds, mammals, and other factors [121]. Among agricultural control methods, cultural practices, especially the use of resistant varieties, are among the most effective strategies [120]. Thanks to breeding and biotechnological advancements, the development of such varieties has become possible [120]. Increasing global population and changing environmental conditions have made it imperative to obtain higher yields and better-quality products per unit area employed [122, 123]. Biotechnological research has advanced significantly with the dual aim of improving product quality and yield while enhancing plant resistance to environmental stressors, particularly diseases and pests, through the application of both classical breeding methods and modern biotechnological techniques [122, 123]. In particular, modern plant breeding methods are now being applied alongside conventional breeding to improve pest resistance, and modern breeding facilitates the identification of traits that contribute to superior performance in distant environments through tissue culture, somatic hybridization, embryo rescue, molecular markers, gene mapping, and gene transfer [124]. A key advantage of modern methods is the ability to transfer genes between unrelated species, and as a result, a gene obtained from a plant, animal, or microorganism can be transferred to a completely different organism and function as a natural component of the recipient genome [122, 123]. While transgenic plants developed for insect resistance offer benefits such as reduced pesticide use, environmental protection, lower production costs, and increased profitability for farmers, they also pose certain concerns [125]. The first studies on plant resistance to insects were in 1831, when Lindley identified resistance in the "Winter Majetin" apple cultivar against *Eriosoma lanigerum* Hausm. (Homoptera: Pemphigidae) [126]. Later, resistance breeding against *Mayetiola destructor* Say (Diptera: Cecidomyiidae), which caused significant damage to cereal crops and was introduced from Europe to the United States in 1788, began with the use of the "Underkill" variety [126].

Use of *Bacillus thuringiensis* (Bt) in Insect Resistance: Over the past four decades, hundreds of insect species have developed resistance to one or more control methods [127]. More than 500 pest species have been identified as resistant to at least one insecticide [128]. On a global scale, 2009 and 2010 data show that 61% of transgenic crop cultivation areas featured herbicide-tolerant varieties, while 17% featured insect-resistant varieties [129]. Soybean, maize, cotton, and canola constitute 99% of all commercially cultivated transgenic crops [129]. In countries like Australia, China, Spain, and the United States, Bt crops have been used for over 10 years against six major lepidopteran species, with ongoing monitoring to detect any resistance [130]. Bt crops are expected to significantly reduce pesticide usage through reduced application frequency and smaller treatment areas [131, 132]. For example, in Australia, Bt cotton reduced insecticide applications by 59% and active ingredient use by 44% between 1996 and 2004 and between 2002 and 2006, cotton varieties carrying two Bt genes reduced insecticide use by 85% and active ingredient use by 65–75% [133]. In China, Bt cotton reduced insecticide usage by 59–66% and total insecticide quantity by 61–80% [133]. Bt crops were found to have no direct toxic effects on predators and parasitoids; adverse effects on natural enemies are observed only when they fed on Bt-sensitive prey or hosts [134–136].

Use of Different Genes for Insect Resistance: The Arcelin gene, responsible for producing the Arcelin protein, has been used to develop insect-resistant plants, for example, resistance to *Zabrotes subfasciatus*, a major storage pest of beans, has been achieved with Arcelin1 and Arcelin4, when added to artificial diets at concentrations exceeding 7% of dry weight, showed toxic effects on *Zabrotes subfasciatus* larvae by inhibiting protein digestion and essential amino acid synthesis [137, 138]. A trypsin inhibitor (CpTI) is shown to be effective against multiple insect pests, including species from the orders Coleoptera, Lepidoptera, and Orthoptera [139].

7. Conclusions

Plant-parasitic nematodes, bacterial pathogens, insect pests, viroids, and viruses collectively impose significant threats to global agriculture, challenging crop productivity and food security. Advances in molecular biology and biotechnology have transformed resistance breeding, enabling the development of crops with enhanced, durable, and broad-spectrum resistance. Integration of traditional breeding with cutting-edge tools such as RNA interference, CRISPR/Cas gene editing, and genomic selection has accelerated the identification and deployment of resistance genes across diverse crop species. Despite these advancements, the rapid evolution of pathogens and pests, coupled with environmental influences and complex host–pathogen interactions, continues to undermine resistance durability. This book chapter reviews plant resistance against plant pests and pathogens. Notably, the resistance and the multiplexing capabilities of RNAi and CRISPR systems offer promising strategies to overcome these challenges. Furthermore, molecular breeding approaches, including genome-wide association studies and multi-omics data integration, are essential for elucidating underlying resistance mechanisms and refining breeding programmes. Moving forward, sustainable disease and pest management will rely on durable resistance, supported by comprehensive genome analyses. Continued novelty, informed by a thorough understanding of host–pathogen co-evolution and molecular interactions, is essential. Ultimately, the convergence of classical genetics, molecular technologies, and precision breeding holds the key to developing resistant crops, defense agricultural productivity, and ensuring global food security in the face of evolving biotic stresses.

Author contribution

The contributions to this work are as follows: Resistance against nematodes in plants (by R. Bozbuga); Resistance and breeding strategies against viral pathogens (by B.B. Arpacı, B. Özgören and Ü.H. Erol); Resistance and breeding strategies against viroids (by P.G. Guler); Resistance mechanisms against insects in plants (by D. Kahya and P.A. Kara); and Resistance and breeding strategies against bacterial pathogens (by H.N. Yildiz) have been authored in this book chapter.

Conflict of interest

The authors declare no conflict of interest.

Author details

Refik Bozbuga[1*], Pinar Arıdıcı Kara[2], Hatice Nilufer Yıldız[2], Pakize Gök Güler[2], Ümit Haydar Erol[3], Burak Özgören[3], Doğancan Kahya[1] and Bekir Bülent Arpacı[3]

1 Department of Plant Protection, Faculty of Agriculture, Eskisehir Osmangazi University, Eskisehir, Turkiye

2 Biological Control Research Institute, Yureğir, Adana, Turkiye

3 Department of Horticulture, Faculty of Agriculture, Kilis 7 Aralık University, Kilis, Turkiye

*Address all correspondence to: refikbozbuga@gmail.com

References

[1] Boag B, Yeates GW. Soil nematode biodiversity in terrestrial ecosystems. Biodiversity and Conservation. 1998;7:617-630. DOI: 10.1023/A:1008852301349

[2] Decraemer W, Hunt DJ. Structure and classification. In: Perry RN, Moens M, editors. Plant Nematology. London: CABI Publishing; 2006. pp. 4-32. DOI: 10.1079/9781845930561.0003

[3] Karakaş US, Peçen A, Kepenekçi İ, Çetintaş R. Determination of plant parasitic nematode species of Tylenchida (Nematoda) in apple (*Pyrus malus* L.) and walnut (*Juglans regia* L.) orchards in Bingöl Province of Turkey. KSÜ Tarım Doğa Derg. 2021;**24**:554-560. DOI: 10.18016/ksutarimdoga.vi.750286

[4] Uludamar EBK, Yıldız Ş, Imren M, Öcal A, Elekçioğlu İH. Occurrence of plant parasitic nematode species in important crops in the Southeast Anatolia region of Turkey. Turkish Journal of Entomology. 2018;**42**:63-74. DOI: 10.16970/entoted.359616

[5] Warmerdam S, Sterken MG, van Schaik C, Oortwijn MEP, Sukarta OCA, Lozano-Torres JL, et al. Genome-wide association mapping of the architecture of susceptibility to the root-knot nematode *Meloidogyne incognita* in *Arabidopsis thaliana*. The New Phytologist. 2018;**218**:724-737. DOI: 10.1111/nph.15034

[6] Williamson MW, Hussey RS. Nematode pathogenesis and resistance in plants. The Plant Cell. 1996;**8**:1735-1745. DOI: 10.1105/tpc.8.10.1735

[7] Sato K, Kadota Y, Shirasu K. Plant immune responses to parasitic nematodes. Frontiers in Plant Science. 2019;**10**:1165. DOI: 10.3389/fpls.2019.01165

[8] Escobar C, Barcala M, Cabrera J, Fenoll C. Overview of root-knot nematodes and giant cells. In: Escobar C, Fenoll C, editors. Plant Nematode Interactions: A View on Compatible Interrelationships. Oxford: Elsevier; 2015. pp. 1-32. DOI: 10.1016/bs.abr.2015.01.001

[9] Abad P, Castagnone-Sereno P, Rosso MN, Engler JA, Favery B. Invasion, feeding and development. In: Perry RN, Moens M, Starr JL, editors. Root-Knot Nematodes. Cambridge: CABI International Press; 2009. pp. 163-181. DOI: 10.1079/9781845934927.0163

[10] Mejias J, Bazin J, Truong NM, Chen Y, Marteu N, Bouteiller N, et al. The root-knot nematode effector MiEFF18 interacts with the plant core spliceosomal protein SmD1 required for giant cell formation. The New Phytologist. 2021;**229**:3408-3423. DOI: 10.1111/nph.17089

[11] Quentin M, Abad P, Favery B. Plant parasitic nematode effectors target host defense and nuclear functions to establish feeding cells. Frontiers in Plant Science. 2013;**4**:53. DOI: 10.3389/fpls.2013.00053

[12] Ali MA, Azeem F, Abbas A, Joyia FA, Li H, Dababat AA. Transgenic strategies for enhancement of nematode resistance in plants. Frontiers in Plant Science. 2017;**8**:750. DOI: 10.3389/fpls.2017.00750

[13] Hou S, Liu Z, Shen H, Wu D. Damage-associated molecular pattern-triggered immunity in plants. Frontiers in Plant Science. 2019;**10**:646. DOI: 10.3389/fpls.2019.00646

[14] Boutrot F, Zipfel C. Function, discovery, and exploitation of plant pattern recognition receptors for broad-spectrum disease resistance. Annual Review of Phytopathology. 2017;**55**:257-286. DOI: 10.1146/annurev-phyto-080614-120106

[15] Fuller VL, Lilley CJ, Urwin PE. Nematode resistance. The New Phytologist. 2008;**180**:27-44. DOI: 10.1111/j.1469-8137.2008.02508.x

[16] Kaloshian I, Desmond OJ, Atamian HS. Disease resistance-genes and defense responses during incompatible interactions. In: Jones J, Gheysen G, Fenoll C, editors. Genomics and Molecular Genetics of Plant-Nematode Interactions. Dordrecht: Springer; 2011. pp. 309-324. DOI: 10.1007/978-94-007-0434-3_15

[17] Williamson VM, Kumar A. Nematode resistance in plants: The battle underground. Trends in Genetics. 2006;**22**:396-403. DOI: 10.1016/j.tig.2006.05.003

[18] Williamson VM. Root-knot nematode resistance genes in tomato and their potential for future use. Annual Review of Phytopathology. 1998;**36**:277-293. DOI: 10.1146/annurev.phyto.36.1.277

[19] Cook R. Genetic resistance to nematodes: Where is it useful? Australasian Plant Pathology. 2004;**33**:139-150. DOI: 10.1071/AP04035

[20] Savary S, Ficke A, Aubertot JN, Hollier C. Crop losses due to diseases and their implications for global food production losses and food security. Food Security. 2012;**4**:519-537. DOI: 10.1007/s00203-017-1426-6

[21] Hirooka T, Ishii H. Chemical control of plant diseases. Journal of General Plant Pathology. 2013;**79**:390-401. DOI: 10.1007/s10327-013-0470-6

[22] McManus PS, Stockwell VO, Sundin GW, Jones AL. Antibiotic use in plant agriculture. Annual Review of Phytopathology. 2002;**40**:443-465. DOI: 10.1146/annurev.phyto.40.120301.093927

[23] Cooksey DA. Genetics of bactericide resistance in plant pathogenic bacteria. Annual Review of Phytopathology. 1990;**28**:201-219. DOI: 10.1146/annurev.py.28.090190.001221

[24] Behlau F, Hong JC, Jones JB, Graham JH. Evidence for acquisition of copper resistance genes from different sources in citrus-associated xanthomonads. Phytopathology. 2013;**103**:409-418. DOI: 10.1094/PHYTO-06-12-0134-R

[25] Moller WJ, Schroth MN, Thomson SV. The scenario of fire blight and streptomycin resistance. Plant Disease. 1981;**65**:563-568. DOI: 10.1094/PD-65-563

[26] Sundin GW, Bender CL. Expression of the strA-strB streptomycin resistance genes in *pseudomonas syringae* and *Xanthomonas campestris* and characterization of IS6100 in *Xanthomonas campestris*. Applied and Environmental Microbiology. 1995;**61**:2891-2897. DOI: 10.1128/aem.61.8.2891-2897.1995

[27] Qaim M. Role of new plant breeding technologies for food security and sustainable agricultural development. Applied Economic Perspectives and Policy. 2020;**42**(2):129-150. DOI: 10.1002/aepp.13044

[28] Mookiah S, Sivasubramaniam B, Thangaraj T, Govindaraj S. Host Plant Resistance Molecular Approaches for

Sustainable Insect Pest Management. Singapore: Springer; 2021. pp. 1-56. DOI: 10.1007/978-981-16-3591-5_1

[29] Bharadwaj DN. Sustainable agriculture and plant breeding. In: Advances Plant Breed Strategies Agron Abiotic and Biotic Stress Traits. Cham: Springer; 2016. pp. 3-34. DOI: 10.1007/978-3-319-22518-0

[30] Mourgues F, Brisset MN, Chevreau E. Strategies to improve plant resistance to bacterial diseases through genetic engineering. Trends in Biotechnology. 1998;**16**(5):203-210. DOI: 10.1016/S0167-7799(98)01189-5

[31] Postel S, Kemmerling B. Plant systems for recognition of pathogen-associated molecular patterns. Seminars in Cell & Developmental Biology. 2009;**20**(9):1025-1031. DOI: 10.1016/j.semcdb.2009.06.002

[32] Jones JD, Dangl JL. The plant immune system. Nature. 2006;**444**(7117):323-329. DOI: 10.1038/nature05286

[33] Alfano JR, Collmer A. Type III secretion system effector proteins: Double agents in bacterial disease and plant defense. Annual Review of Phytopathology. 2004;**42**:385-414. DOI: 10.1146/annurev.phyto.42.040103.110731

[34] Kombrink E, Schmelzer E. The hypersensitive response and its role in local and systemic disease resistance. European Journal of Plant Pathology. 2001;**107**:69-78. DOI: 10.1023/A:1008736629717

[35] Morel JB, Dangl JL. The hypersensitive response and the induction of cell death in plants. Cell Death and Differentiation. 1997;**4**(8):671-683. DOI: 10.1038/sj.cdd.4400309

[36] Yang S, Peng Q, San Francisco M, Wang Y, Zeng Q, Yang CH. Type III secretion system genes of *Dickeya dadantii* 3937 are induced by plant phenolic acids. PLoS One. 2008;**3**:e2973. DOI: 10.1371/annotation/91170966-226f-4678-999e-22f2c4a6bb8d

[37] Kang YJ, Kim KH, Shim S, Yoon MY, Sun S, Kim MY, et al. Genome-wide mapping of NBS-LRR genes and their association with disease resistance in soybean. BMC Plant Biology. 2012;**12**:139. DOI: 10.1186/1471-2229-12-139

[38] Leach JE, Vera Cruz CM, Bai J, Leung H. Pathogen fitness penalty as a predictor of durability of disease resistance genes. Annual Review of Phytopathology. 2001;**39**:187-224. DOI: 10.1146/annurev.phyto.39.1.187

[39] Galis I, Gaquerel E, Pandey SP, Baldwin IT. Molecular mechanisms underlying plant memory in JA-mediated defence responses. Plant, Cell & Environment. 2009;**32**(6):617-627. DOI: 10.1111/j.1365-3040.2008.01862.x

[40] Kachroo P, Kachroo A. The roles of salicylic acid and jasmonic acid in plant immunity. Molecular Plant Immunity. (Hoboken, NJ, USA: Wiley). 2012:55-79. DOI: 10.1002/9781118481431

[41] Cock J, Gitterle T, Salazar M, Rye M. Breeding for disease resistance of penaeid shrimps. Aquaculture. 2009;**286**(1-2):1-11. DOI: 10.1016/j.aquaculture.2008.09.011

[42] Borrelli VM, Brambilla V, Rogowsky P, Marocco A, Lanubile A. The enhancement of plant disease resistance using CRISPR/Cas9 technology. Frontiers in Plant Science. 2018;**9**:1245. DOI: 10.3389/fpls.2018.01245

[43] Boubakri H. Recent progress in CRISPR/Cas9-based genome editing for enhancing plant disease resistance.

Gene. 2023;**866**:147334. DOI: 10.1016/j.gene.2023.147334

[44] Chen X, Liu P, Mei L, He X, Chen L, Liu H, et al. Xa7, a new executor R gene that confers durable and broad-spectrum resistance to bacterial blight disease in rice. Plant Communications. 2021;**2**(3): 1-14. DOI: 10.1016/j.xplc.2021.100143

[45] Zafar K, Khan MZ, Amin I, Mukhtar Z, Yasmin S, Arif M, et al. Precise CRISPR-Cas9 mediated genome editing in super basmati rice for resistance against bacterial blight by targeting the major susceptibility gene. Frontiers in Plant Science. 2020;**11**:575. DOI: 10.3389/fpls.2020.00575

[46] Teper D, White FF, Wang N. The dynamic transcription activator-like effector family of *Xanthomonas*. Phytopathology. 2023;**113**(4):651-666. DOI: 10.1094/PHYTO-10-22-0365-KD

[47] Boch J, Bonas U, Lahaye T. TAL effectors–pathogen strategies and plant resistance engineering. The New Phytologist. 2014;**204**:823-832. DOI: 10.1111/nph.13015

[48] Janse JD. Bacterial diseases that may or do emerge, with (possible) economic damage for Europe and the Mediterranean basin: Notes on epidemiology, risks, prevention and management on first occurrence. Journal of Plant Pathology. 2012;**94**:S5-S29. Available from: http://www.jstor.org/stable/45156305

[49] Lopes CA, Boiteux LS. Breeding for resistance to bacterial diseases. In: Plant Breed for Biotic Stress Resistance. Berlin, Heidelberg: Springer; 2012. pp. 37-55. DOI: 10.1007/978-3-642-33087-2_3

[50] Coyne DP, Schuster ML. Breeding and genetic studies of tolerance to several bean (*Phaseolus vulgaris* L.) bacterial pathogens. Euphytica. 1974;**23**(3):651-656. DOI: 10.1007/BF00022486

[51] Chattopadhyay A, Seth T, Dutta S, Ghosh PP, Chattopadhyay SB, Majumder D, et al. Breeding eggplant for higher productivity and bacterial wilt tolerance. International Journal of Vegetable Science. 2012;**18**(4):376-392. DOI: 10.1080/19315260.2011.653625

[52] Emeriewen OF, Wöhner T, Flachowsky H, Peil A. *Malus* hosts–*Erwinia amylovora* interactions: Strain pathogenicity and resistance mechanisms. Frontiers in Plant Science. 2019;**10**:551. DOI: 10.3389/fpls.2019.00551

[53] Sánchez-Martín J, Keller B. Contribution of recent technological advances to future resistance breeding. Theoretical and Applied Genetics. 2019;**132**(3):713-732. DOI: 10.1007/s00122-019-03297-1

[54] Martínez JL, Baquero F. Interactions among strategies associated with bacterial infection: Pathogenicity, epidemicity, and antibiotic resistance. Clinical Microbiology Reviews. 2002;**15**(4):647-679. DOI: 10.1128/cmr.15.4.647-679.2002

[55] Blum A, Jordan WR. Breeding crop varieties for stress environments. Critical Reviews in Plant Sciences. 1985;**2**(3):199-238. DOI: 10.1080/07352688509382196

[56] Isbell F, Craven D, Connolly J, Loreau M, Schmid B, Beierkuhnlein C, et al. Biodiversity increases the resistance of ecosystem productivity to climate extremes. Nature. 2015;**526**(7574):574-577. DOI: 10.1038/nature15374

[57] Flores R, Minoia S, Carbonell A, Gisel A, Delgado S, López-Carrasco A, et al. Viroids, the simplest RNA replicons: How they manipulate their hosts for

being propagated and how their hosts react for containing the infection. Virus Research. 2015;**209**:136-145. DOI: 10.1016/j.virusres.2015.02.027

[58] Navarro B, Flores R, Di Serio F. Advances in viroid-host interactions. Annual Review of Virology. 2021;**8**:305-325. DOI: 10.1146/annurev-virology-091919-092331

[59] Ortola B, Daròs JA. Viroids: Non-coding circular RNAs able to autonomously replicate and infect higher plants. Biology. 2023;**12**:172. DOI: 10.3390/biology12020172

[60] Diener TO. Viroids: The smallest known agents of infectious disease. Annual Review of Microbiology. 1974;**28**:23-40. DOI: 10.1146/annurev.mi.28.100174.000323

[61] Navarro B, Turina M. Viroid and viroid-like elements in plants and plant-associated microbiota: A new layer of biodiversity for plant holobionts. The New Phytologist. 2024;**244**(4):1216-1222. DOI: 10.1111/nph.20156

[62] Minoia S, Carbonell A, Di Serio F, Gisel A, Carrington JC, Navarro B, et al. Specific argonautes selectively bind small RNAs derived from potato spindle tuber viroid and attenuate viroid accumulation in vivo. Journal of Virology. 2014;**88**:11933-11945. DOI: 10.1128/jvi.01404-14

[63] Flores R, Di Serio F, Navarro B, Duran-Vila N, Owens RA. Viroids and viroid diseases of plants. Studies in Viral Ecology. 2011;**1**:267-296. DOI: 10.1002/9781118025666.ch12

[64] Diener TO. Discovering viroids—A personal perspective. Nature Reviews. Microbiology. 2003;**1**:75-80. DOI: 10.1038/nrmicro736

[65] Ding B, Itaya A. Viroid: A useful model for studying the basic principles of infection and RNA biology. Molecular Plant-Microbe Interactions. 2007;**20**:7-20. DOI: 10.1094/mpmi-20-0007

[66] Diener TO. Potato spindle tuber "virus": IV. A replicating, low molecular weight RNA. Virology. 1971;**45**:411-428. DOI: 10.1016/0042-6822(71)90342-4

[67] Kovalskaya N, Hammond RW. Molecular biology of viroid–host interactions and disease control strategies. Plant Science. 2014;**228**:48-60. DOI: 10.1016/j.plantsci.2014.05.006

[68] Diener TO. Circular RNAs – Relics of precellular evolution. Proceedings of the National Academy of Sciences of the United States of America. 1989;**86**:9370-9374. DOI: 10.1073/pnas.86.23.9370

[69] Diener TO. Origin and evolution of viroids and viroid-like satellite RNAs. Virus Genes. 1996;**11**:119-131. DOI: 10.1007/BF01728653

[70] Diener TO. The viroid: Biological oddity or evolutionary fossil? Advances in Virus Research. 2001;**57**:137-184. DOI: 10.1016/S0065-3527(01)57003-7

[71] Naoi T, Li S, Sano T. Resistance and tolerance to viroid infection. Biology and Life Sciences. Preprints. 2022. DOI: 10.20944/preprints202211.0169.v1

[72] Lai MM. The molecular biology of hepatitis delta virus. Annual Review of Biochemistry. 1995;**64**:259-286. DOI: 10.1146/annurev.bi.64.070195.001355

[73] Taylor JM. Hepatitis delta virus. Intervirology. 1999;**42**:173-178. DOI: 10.1159/000024977

[74] Flores R, Delgado S, Rodio ME, Ambrós S, Hernández C, Di

Serio F. Peach latent mosaic viroid: Not so latent. Molecular Plant Pathology. 2006;**7**:209-221. DOI: 10.1111/j.1364-3703.2006.00332.x

[75] Owens RA, Hammond R. Viroid pathogenicity: One process, many faces. Viruses. 2009;**1**:298-316. DOI: 10.3390/v1020298

[76] Verhoeven JTJ, Jansen CCC, Willemen TM, Kox LFF, Owens RA, Roenhorst JW. Natural infections of tomato by citrus exocortis viroid, Columnea latent viroid, potato spindle tuber viroid, and tomato chlorotic dwarf viroid. European Journal of Plant Pathology. 2004;**110**:823-831. DOI: 10.1007/s10658-004-2493-5

[77] Verhoeven JT, Humer L, Marn MV, Plessk IM, Roenhorst JW. Mechanical transmission of potato spindle tuber viroid between plants of *Brugmansia suaveolens*, *Solanum jasminoides* and potatoes and tomatoes. European Journal of Plant Pathology. 2010;**128**:417-421. DOI: 10.1007/s10658-010-9675-0

[78] Matsushita Y, Aoki K, Sumitomo K. Selection and inheritance of resistance to chrysanthemum stunt viroid. Crop Protection. 2012;**35**:1-4. DOI: 10.1016/j.cropro.2011.12.009

[79] Nabeshima T, Hosokawa M, Yano S, Ohishi K, Doi M. Screening of *Chrysanthemum* cultivars with resistance to *Chrysanthemum* stunt viroid. Journal of the Japanese Society for Horticultural Science. 2012;**81**:285. DOI: 10.2503/jjshs1.81.285

[80] Lizárraga RE, Salazar SF, Roca WM, Schilde-Rentschler L. Elimination of potato spindle tuber viroid by low temperature and meristem culture. Phytopathology. 1980;**70**:754-755. DOI: 10.1094/phyto-70-754

[81] Carbonell A, Flores R, Gago S. Hammerhead ribozymes against virus and viroid RNAs. In: Erdmann VA, Barciszewski J, editors. From Nucleic Acid Sequences to Molecular Medicine. RNA Technologies. Berlin Heidelberg: Springer-Verlag; 2012. pp. 411-427. DOI: 10.1007/978-3-642-27426-8_16

[82] Bari R, Jones JDG. Role of plant hormones in plant defence responses. Plant Molecular Biology. 2009;**69**:473-488. DOI: 10.1007/s11103-008-9435-0

[83] Campos L, Granell P, Tárraga S, López-Gresa P, Conejero V, Bellés JM, et al. Salicylic acid and gentisic acid induce RNA silencing-related genes and plant resistance to RNA pathogens. Plant Physiology and Biochemistry. 2014;**77**:35-43. DOI: 10.1016/j.plaphy.2014.01.016

[84] Li S, Zhang Z, Zhou C, Li S. RNA-dependent RNA polymerase 1 delays the accumulation of viroids in infected plants. Molecular Plant Pathology. 2021;**22**(10):1195-1208. DOI: 10.1111/mpp.13104

[85] Itaya A, Zhong X, Bundschuh R, Qi Y, Takeda YR, Harris AR, et al. A structured viroid RNA serves as a substrate for dicerlike cleavage to produce biologically active small RNAs but is resistant to RNA-induced silencing complex-mediated degradation. Journal of Virology. 2007;**81**:2980-2994. DOI: 10.1128/jvi.02339-06

[86] Matousek J, Schröder ARW, Trnená L, Reimers M, Baumstark T, Dedic P, et al. Inhibition of viroid infection by antisense RNA expression in transgenic plants. Biological Chemistry Hoppe-Seyler. 1994;**375**:765-777. DOI: 10.1515/bchm3.1994.375.11.765

[87] Yang X, Yie Y, Zhu F, Liu Y, Kang L, Wang X, et al. Ribozyme-mediated high resistance against potato spindle

tuber viroid in transgenic potatoes. Proceedings of the National Academy of Sciences of the United States of America. 1997;**94**:4861-4865. DOI: 10.1073/pnas.94.10.4861

[88] Schwind N, Zwiebel M, Itaya A, Ding B, Wang MB, Krczal G, et al. RNAi-mediated resistance to potato spindle tuber viroid in transgenic tomato expressing a viroid hairpin RNA construct. Molecular Plant Pathology. 2009;**10**:459-469. DOI: 10.1111/j.1364-3703.2009.00546.x

[89] Ogawa T, Toguri T, Kudoh H, Okamura M, Momma T, Yoshioka M, et al. Double-stranded RNA-specific ribonuclease confers tolerance against *Chrysanthemum* stunt viroid and tomato spotted wilt virus in transgenic *Chrysanthemum* plants. Breeding Science. 2005;**55**:49-55. DOI: 10.1270/jsbbs.55.49

[90] Sano T, Nagayama A, Ogawa T, Ishida I, Okada Y. Transgenic potato expressing a double-stranded RNA-specific ribonuclease is resistant to potato spindle tuber viroid. Nature Biotechnology. 1997;**15**:1290-1294. DOI: 10.1038/nbt1197-1290

[91] Singh RP, Clark MC. Infectious low-molecular weight ribonucleic acid from tomato. Biochemical and Biophysical Research Communications. 1971;**44**:1077-1083. DOI: 10.1016/S0006-291X(71)80195-X

[92] Navarro B, Gisel A, Rodio ME, Delgado S, Flores R, Di Serio F. Viroids: How to infect a host and cause disease without encoding proteins. Biochimie. 2012;**94**(7):1474-1480. DOI: 10.1016/j.biochi.2012.02.020

[93] Grech-Baran M, Stolt-Bergner P, Mulaudzi-Kabunda N. Extreme resistance to potato virus Y in potato carrying the Rysto gene is mediated by a TIR-NLR immune receptor. Plant Biotechnology Journal. 2020;**18**:655-667. DOI: 10.1111/pbi.13230

[94] Zhang C, Grosic S, Whitham SA, Hill JH. The requirement of multiple defense genes in soybean Rsv1-mediated extreme resistance to soybean mosaic virus. Molecular Plant-Microbe Interactions. 2012;**25**:1307-1313. DOI: 10.1094/MPMI-02-12-0046-R

[95] Valkonen JPT, Pethybridge SJ, Skoglund L. Evidence for utility of the same PCR-based markers for selection of extreme resistance to potato virus Y controlled by Rysto of *Solanum stoloniferum*. The Annals of Applied Biology. 2008;**152**:121-130. DOI: 10.1111/j.1744-7348.2007.00194.x

[96] Huang W, Long J, Xu H. Extreme resistance to potato virus a in potato cultivar Barbara is independently mediated by Ra and Rysto. Plant Disease. 2021;**105**:3344-3348. DOI: 10.1094/PDIS-02-21-0233-SC

[97] Janzac B, Bitton F, Laffaire JB. Phenotype and spectrum of action of the Pvr4 resistance in pepper against potyviruses and selection for virulent variants. Plant Pathology. 2009;**58**:443-449. DOI: 10.1111/j.1365-3059.2008.01992.x

[98] Ghazala W, Varrelmann M. Tobacco rattle virus 29K movement protein is the elicitor of extreme and hypersensitive-like resistance in two cultivars of *Solanum tuberosum*. Molecular Plant-Microbe Interactions. 2007;**20**:1396-1405. DOI: 10.1094/MPMI-20-11-1396

[99] Iwai H, Terao T, Hibi T. A recessive resistance gene of alfalfa to alfalfa mosaic virus and impact of high ambient temperature on virus resistance. Plant Pathology. 1992;**41**:69-75. DOI: 10.1111/j.1365-3059.1992.tb02318.x

[100] Kim BM, Lee SW, Kim KS. The P3 protein of turnip mosaic virus can alone induce hypersensitive response-like cell death in Arabidopsis thaliana. Molecular Plant-Microbe Interactions. 2010;**23**:144-152. DOI: 10.1094/MPMI-23-2-0144

[101] Lovato FA, Muniz MB, Marubayashi JM. The N protein of tomato spotted wilt virus is associated with the induction of programmed cell death in *Capsicum chinense*. Virus Research. 2008;**137**:245-252. DOI: 10.1016/j.virusres.2008.07.020

[102] Rodamilans B, de Oliveira AC, Valli A. The mechanism of resistance of European plum to plum pox virus is linked to NIa protease activity. Plants. 2023;**12**:1609. DOI: 10.3390/plants12081609

[103] Hussain M, Amin I, Naimuddin Khan M. The hypersensitive response to tomato leaf curl New Delhi virus nuclear shuttle protein is inhibited by transcriptional activator protein. Molecular Plant-Microbe Interactions. 2007;**20**:1581-1588. DOI: 10.1094/MPMI-20-12-1581

[104] Huh SU, Kim YJ, Kim SY. *Capsicum annuum* WRKY transcription factor d regulates hypersensitive response upon TMV infection. Plant Science. 2012;**197**:50-58. DOI: 10.1016/j.plantsci.2012.08.013

[105] Zhao L, Wang X, Li Y. TZSV-induced hypersensitive resistance via an auxin-related pathway in pepper. Gene. 2022;**823**:146320. DOI: 10.1016/j.gene.2022.146320

[106] Moury B, Valkonen JPT, Guiraud T. High temperature effects on hypersensitive resistance to tomato spotted wilt virus in pepper. European Journal of Plant Pathology. 1998;**104**:489-498. DOI: 10.1023/A:1008618022144

[107] Takahashi H, Kusano T, Hayashida N. Hypersensitive response in cucumber mosaic virus-inoculated *Arabidopsis thaliana*. The Plant Journal. 1994;**6**:369-377. DOI: 10.1046/j.1365-313X.1994.06030369.x

[108] Tamisier L, Szadkowski M, Girardot G. Concurrent evolution of resistance and tolerance to potato virus Y in *Capsicum annuum* revealed by genome-wide association. Molecular Plant Pathology. 2022;**23**:254-264. DOI: 10.1111/mpp.13157

[109] Lapidot M, Paran I, Ben-Joseph R. Tolerance to cucumber mosaic virus in pepper: Development of advanced breeding lines and evaluation of virus level. Plant Disease. 1997;**81**:185-188. DOI: 10.1094/PDIS.1997.81.2.185

[110] Eldan O, Ofir A, Luria N. Pepper plants harboring L resistance alleles showed tolerance toward manifestations of tomato brown rugose fruit virus disease. Plants. 2022;**11**:2378. DOI: 10.3390/plants11182378

[111] Soler S, Debreczeni DE, Vidal E. A new *Capsicum baccatum* accession shows tolerance to wild-type and resistance-breaking isolates of tomato spotted wilt virus. The Annals of Applied Biology. 2015;**167**:343-353. DOI: 10.1111/aab.12229

[112] Pooggin MM. RNAi-mediated resistance to viruses: A critical assessment of methodologies. Current Opinion in Virology. 2017;**26**:28-35. DOI: 10.1016/j.coviro.2017.07.010

[113] Rodríguez-Hernández AM, Gosalvez B, Sempere RN. Melon RNA interference (RNAi) lines silenced for Cm-eIF4E show broad virus resistance. Molecular Plant Pathology. 2012;**13**:755-763. DOI: 10.1111/j.1364-3703.2012.00785.x

[114] Hameed A, Tahir MN, Asad S. RNAi-mediated simultaneous resistance against three RNA viruses in potato. Molecular Biotechnology. 2017;**59**:73-83. DOI: 10.1007/s12033-017-9995-9

[115] Khatodia S, Bhatotia K, Tuteja N. Development of CRISPR/Cas9 mediated virus resistance in agriculturally important crops. Bioengineered. 2017;**8**:274-279. DOI: 10.1080/21655979.2017.1297347

[116] Tashkandi M, Ali Z, Aljedaani F. Engineering resistance against tomato yellow leaf curl virus via the CRISPR/Cas9 system in tomato. Plant Signaling & Behavior. 2018;**13**:e1525996. DOI: 10.1080/15592324.2018.1525996

[117] Liu H, Soyars CL, Li J. CRISPR/Cas9-mediated resistance to cauliflower mosaic virus. Plant Direct. 2018;**2**:e00047. DOI: 10.1002/pld3.47

[118] Tiwari JK, Tuteja N, Khurana SP. Genome editing (CRISPR-Cas)-mediated virus resistance in potato (*Solanum tuberosum* L.). Molecular Biology Reports. 2022;**49**:12109-12119. DOI: 10.1007/s11033-022-07704-7

[119] Lucioli A, Tavazza R, Baima S. CRISPR-Cas9 targeting of the eIF4E1 gene extends the potato virus Y resistance spectrum of the *Solanum tuberosum* L. cv. Desirée. Frontiers in Microbiology. 2022;**13**:873930. DOI: 10.3389/fmicb.2022.873930

[120] Keçeci M, Baysal Ö, Soysal M, Tekşam İ. Bitkilerde böceklere dayanıklılık mekanizmaları. Derim. 2007;**24**(1):19-31. Available from: https://dergipark.org.tr/en/pub/derim/issue/4562/62663

[121] Ecevit O, Tuncer C. Gen transferi ile böceklere karşı dayanıklı bitki elde etme çalışmaları. Türk Entomol Derg. 1991;**15**(2):117-127. Available from: https://dergipark.org.tr/tr/pub/entoted/issue/5611/75515

[122] Primrose SB. Molecular Biotechnology. Oxford: Blackwell Scientific Publications; 1991. p. 196. ISBN 9780632030538

[123] Sökmen M. Genetik yapisi değiştirilmiş bitkiler ve bitki koruma amaçli kullanimi. Anadolu Tarım Bilimleri Dergisi. 2005;**20**(3):105-109. DOI: 10.7161/anajas.2005.20.3.105-109

[124] Ewing PM, Runck BC, Kono TYJ, Kantar MB. The home field advantage of modern plant breeding. PLoS One. 2019;**14**(12):e0227079. DOI: 10.1371/journal.pone.0227079

[125] Kulikov AM. Genetically modified organisms and risks of their introduction. Russian Journal of Plant Physiology. 2005;**52**(1):99-111. DOI: 10.1007/s11183-005-0015-8

[126] Smith CM. Plant Resistance to Insects. Moscow: Wiley-Interscience Publication; 1989. p. 286. ISBN 9780471849384

[127] MacIntosh SC. Managing the risk of insect resistance to transgenic insect control traits: Practical approaches in local environments. Pest Management Science. 2009;**65**:100-106. DOI: 10.1002/ps.1854

[128] Ali S, Zafar Y, Xianyin Z, Ali GM, Jumin T. Transgenic crops: Current challenges and future perspectives. African Journal of Biotechnology. 2008;**7**(25):4667-4676. DOI: 10.5897/AJB08.004

[129] Anonymous. Who Benefits from GM Crops? Friends of the Earth International; 2010. Available from: http://www.foeeurope.org/GMOs/Who_Benefits/who_benefits_full_report_2010.pdf [Accessed: May 01, 2025]

[130] Tabashnik BE, Gassmann AJ, Crowder DW, Carrière Y. Insect resistance to Bt crops: Evidence versus theory. Nature Biotechnology. 2008;**26**(2):199-202. DOI: 10.1038/nbt1382

[131] Anonymous. Who Benefits from GM Crops? The Rise in Pesticide Use. Friends of the Earth International; 2008. Available from: https://www.foei.org/wp-content/uploads/2014/04/18-foei-who-benefits-report-mr.pdf. [Accessed: May 01, 2025]

[132] Lövei GL. Ecological risks and benefits of transgenic plants. New Zealand Plant Protection. 2001;**54**:93-100. DOI: 10.30843/nzpp.2001.54.3723

[133] Naranjo SE. Impacts of Bt crops on non-target invertebrates and insecticide use patterns. CABI Reviews. 2009;**4**(11):1-23. DOI: 10.1079/PAVSNNR20094011

[134] Head G, Moar W, Eubanks M, Freeman B, Ruberson J, Hagerty A, et al. A multiyear, large-scale comparison of arthropod populations on commercially managed Bt and non-Bt cotton fields. Environmental Entomology. 2005;**34**(5):1257-1266. DOI: 10.1093/ee/34.5.1257

[135] Romeis J, Meissle M, Bigler F. Transgenic crops expressing bacillus thuringiensis toxins and biological control. Nature Biotechnology. 2006;**24**(1):63-71. DOI: 10.1038/nbt1180

[136] Tabashnik BE, Carrière Y. Insect resistance to genetically modified crops. In: Ferry N, Gatehouse ARR, editors. Environmental Impact of Genetically Modified Crops. UK: CABI, Nosworth Way, Wallingford, Oxfordshire; 2009. pp. 74-100. DOI: 10.1079/9781845934095.0074

[137] Osborn TC, Alexander DC, Sun SSM, Cardona C, Bliss FA. Insecticidal activity and lectin homology of arcelin seed protein. Science. 1988;**240**:207-210. DOI: 10.1126/science.240.4849.207

[138] Minney BHP, Gatehouse AMR, Dobie P, Denedy J, Cardona C, Gatehouse JA. Biochemical bases of seed resistance to *Zabrotes subfascinatus* (bean weevil) in *Phaseolus vulgaris* (common bean): A mechanism for arcelin toxicity. Journal of Insect Physiology. 1990;**36**:757-767. DOI: 10.1016/0022-1910(90)90049-L

[139] Tosun M, Akgün İ, Sağsöz S. Böceklere dayanıklılık ıslahında genetik mühendisliğinin kullanımı. Atatürk Üniv Ziraat Fak Derg. 1998;**29**(1):142-152. Available from: https://dergipark.org.tr/en/pub/ataunizfd/issue/3020/41917

Chapter 4

Introgression of Biotic Stress Resistance Genes: Enhancing Crop Protection through Molecular Breeding

Vividha M. Sirsat, Aditi R. Rajane, Pallavi R. Sasane and Bhagyashree P. Bahatkar

Abstract

Biotic stresses present major challenges to global food security, frequently causing significant crop losses. To address these threats, integrating introgression breeding strategies with cutting-edge molecular tools can enhance crop resilience and ensure long-term protection. By leveraging advanced biotechnological methods, these approaches not only strengthen disease and pest resistance but also contribute to sustainable agricultural productivity. As climate change intensifies biotic pressures, the adoption of molecular breeding techniques becomes increasingly essential. This chapter delves into the latest innovations in breeding, highlighting how biotechnology aids in refining the genetic architecture of crops, ultimately fostering improved yield stability and adaptability in changing environments.

Keywords: molecular tools, introgression breeding, crop protection, food security, resistance genes

1. Introduction

The increasing global population is placing unprecedented pressure on the agricultural sector to meet rising food demands. However, this growing need is constrained by finite natural resources and compounded by the adverse impacts of climate change. Shifting climatic conditions are disrupting ecological balance, altering precipitation patterns, and intensifying biotic stress, thereby posing significant challenges to global food security and sustainable crop production [1]. Enhancing and maintaining crop quality under these constraints necessitates a multifaceted approach involving the adoption of advanced agronomic practices and the development of high-performing crop varieties. These varieties must exhibit not only superior yield potential but also strong market acceptability, resilience to environmental stress, and improved nutritional value. Balancing productivity with quality and stress resistance is essential for the long-term sustainability of agricultural systems. Among

Figure 1.
Impact of biotic stress on field crop.

the major threats to crop productivity, biotic stress caused by pathogens, pests, and other harmful organisms has become increasingly severe due to changing environmental conditions (**Figure 1**). These biological stress factors lead to significant yield losses and compromise food quality and availability. Effective management of biotic stress is therefore critical to ensure stable crop performance and secure food supplies [2]. Crops are exposed to diverse biotic stressors throughout their growth cycle, with varying degrees of intensity, duration, and impact. This necessitates continuous monitoring and adaptive management practices by growers. A central strategy for mitigating the effects of biotic stress involves the development and deployment of resistant crop cultivars. In recent decades, traditional breeding methods have been employed to develop crops resistant to biotic stresses; however, these approaches are often time-consuming and lack precision. The increasing need for more effective solutions has driven the adoption of molecular breeding, which offers faster and more targeted crop improvement. These modern techniques strengthen resistance to harmful pathogens, thereby enhancing overall crop resilience [3]. This chapter highlights significant examples of crop improvement achieved through modern breeding approaches, emphasizing their role in managing biotic stresses. The case studies showcased underscore the positive impact of these techniques on enhancing productivity and ensuring food security. With ongoing advancements, modern breeding continues to transform agricultural practices, promoting more resilient and sustainable farming systems. Recognizing the importance of these innovations is key to meeting the growing challenges of global agriculture—an idea that will be further explored in the sections ahead.

2. Advances in molecular breeding for crop improvement

Molecular breeding represents a significant advancement in modern plant breeding, leveraging molecular tools to enhance specific desirable traits. It centers on identifying and selecting plants based on genetic markers associated with targeted characteristics, enabling precise and efficient selection at the DNA level. A fundamental aspect of molecular breeding is the analysis of plant genotypic traits, which involves examining the genetic diversity within and between plant populations. This genetic variation is essential, as it serves as the foundation for crop improvement allowing breeders to develop superior plant varieties with increased yields, improved disease resistance, and greater tolerance to environmental stresses [4]. Molecular markers play a crucial role in simplifying the identification of traits responsible for stable resistance in crops. They significantly improve the selection process, especially in situations where biotic stresses limit the effectiveness of conventional breeding. Through the use of these markers, breeders can more rapidly develop robust, high-performing crop varieties with strong resistance to pests and diseases, helping to ensure consistent productivity, even under adverse environmental conditions [5]. Molecular breeding utilizes a variety of advanced techniques, such as Marker-Assisted Selection (MAS), Gene Pyramiding, Marker-Assisted Backcrossing (MABC), Marker-Assisted Recurrent Selection (MARS), Genome-Wide Association Studies (GWAS), and Genomic-Assisted Breeding (GAB), to explore the genetic foundations of plant responses to various stresses. Recently, cutting-edge approaches like Next-Generation Sequencing (NGS), Multi-parent Advanced Generation Intercross (MAGIC), Nested Association Mapping (NAM), and Structured Association (SA) have further enhanced the linkage between genetic markers and traits of interest. These methods offer significant advantages over traditional breeding, particularly in terms of speed, accuracy, and efficiency. While conventional breeding often relies on multiple generations of phenotypic selection and extensive crossbreeding, molecular approaches enable the rapid and precise identification of desirable genetic traits. This will accelerate the incorporation of beneficial traits into crop genomes, streamlining the breeding process. The following section presents key examples of major crops that have been successfully enhanced through the application of molecular breeding techniques.

2.1 Resistance to biotic stress in major field crops

Cereal crops play a vital role in ensuring a stable global food supply, particularly as the world's population continues to grow. While existing varieties have demonstrated strong productivity and yield across various regions, there remains a pressing need to develop improved cultivars capable of withstanding the challenges posed by climate change. Biotic stresses, including infestations by insect pests and outbreaks of diseases, have adversely affected key traits in major cereals, such as wheat, rice, maize, barley, and sorghum. These impacts have led to reductions in both the quantity and quality of harvests. In response, it has become increasingly important to adopt advanced breeding techniques that can accelerate the development of resilient cereal crops equipped with the traits needed to perform well under shifting environmental conditions [6]. Modern biotechnology techniques are crucial for enhancing cereal crops' resistance to biotic stresses. These methods enable researchers to conduct more efficient studies, with scientific reports highlighting their significant value in the improvement of cereal crops. Wheat cultivation faces substantial challenges from diseases, such as leaf rust, yellow rust, and barley yellow dwarf virus, which result in considerable yield losses and threaten global wheat production. These pathogens have impacted vast areas of farmland worldwide, resulting in lower yields and significant crop damage. To address these issues, molecular breeding strategies have been increasingly adopted to improve resistance in wheat cultivars [7]. Notably, the *Lr24* and *Lr28* genes were successfully introgressed into newly developed wheat lines using a marker-assisted backcrossing (MABC) approach, providing enhanced resistance against leaf rust and contributing to more resilient and productive cultivars. Similarly, *Lr19* and *Lr28* genes were deployed through marker-assisted selection (MAS) for effective management of the same disease. In addition, a gene pyramiding strategy was implemented to develop cultivars with multiple resistance traits. This included the combination of *Gn2* and *Gn4* genes for aphid tolerance, along with *CreX* and *CreY* genes for resistance against cyst nematodes, contributing to the development of improved, resilient wheat cultivars [8–11]. Genes conferring resistance to stripe rust were successfully introduced using marker-assisted selection (MAS), enhancing the crop's resilience against this major fungal disease. Likewise, the *PmXQ-0508* gene, which provides resistance to powdery mildew, was introgressed into various wheat genotypes across different breeding populations [12, 13]. Meanwhile, in rice, some major threats, such as gall midge, blast, brown planthopper, bacterial blight, and rice sucker brown planthopper, can severely damage the crop from the early vegetative stage through to the reproductive phase, posing a serious risk to rice production [14]. To combat these threats, a marker-assisted backcrossing (MABC) approach was employed to introgress key resistance genes, such as *Pi9, xa13, Xa21, Gm1, Gm4,* and *Gm8*, into elite cultivars. Additionally, genes like *Xa21, Rc7, Cry1A, Bt,* and *Cry14c* were introgressed using MAS to enhance resistance against various bacterial diseases [14–16]. In rice, a marker-assisted foreground selection approach was utilized to introgress specific *Gm* genes for resistance against gall midge infestation in various breeding populations. Similarly, genes such as *Pi9, Pi47, Pi48*, and *Pi49* for blast resistance, along with genes like *Bph14* and *Bph15* for resistance to brown planthopper, were incorporated using marker-assisted selection (MAS) strategies. In maize, the marker-assisted backcrossing (MABC) approach was employed to introgress the *RppQ* gene, aiming to enhance resistance against both Turcicum Leaf Blight and Polysora Rust [17, 18]. Maize production remains vulnerable to several major diseases, including Maize Rough Dwarf Disease (MRDD), Head Smut, Crimson Leaf Disease, and Maize Chlorotic Mottle Virus, all of which contribute to significant yield losses in many regions around the world. In response, the application of Genome-Wide Association Studies (GWAS) has become increasingly important for identifying genetic traits linked to resistance against these biotic stresses. GWAS has emerged as a powerful and effective tool for uncovering key resistance genes, enabling the development of more resilient maize cultivars and improving overall crop performance. Quantitative trait loci (QTLs) *qHSR1* and *qMrdd8* were identified as being associated with resistance to head smut and maize rough dwarf disease, respectively, through a marker-assisted introgression approach [19, 20]. Similarly, the QTL *qRgls1.06* was identified in maize for resistance against gray leaf spot disease. In sorghum, three QTLs were identified that confer resistance to shoot fly, contributing to improved pest resistance in breeding programs [21]. The identification of QTLs through advanced breeding approaches has become a valuable tool in cereal crop improvement, enabling the discovery of marker-trait associations linked to resistance against various biotic stresses. Genome-Wide Association Studies (GWAS) were employed in maize to identify the *AmZRR16* QTL, which confers resistance to fungal diseases [22]. Similarly, the identified QTL *qRgls1.06* was responsible to confer resistance against gray leaf spot in maize [23]. In barley, the multi-parent advanced generation intercross (MAGIC) population approach led to the identification of key QTLs—*Qsc_3H_1, Qsc_5H_1,* and *Qsc_3H_2*—associated with resistance to barley scald [24]. Additionally, GWAS facilitated the discovery of *QRptt.2H-132.15* and *QPtt.6H-54-55* QTLs linked to resistance against net blotch in barley [25]. In wheat, several *QPm* QTLs were identified as contributing to powdery mildew resistance [26]. The application of advanced breeding techniques offers a promising solution, as they can accelerate the development of resilient cereal varieties capable of withstanding these biotic stresses amid future climate change conditions.

2.2 Resistance to biotic stress in legume crops

Legume crops serve as a vital source of energy and nutrition for both humans and livestock. Their high protein content makes them especially valuable in combating malnutrition and enhancing daily diets. In addition to their nutritional benefits, legumes contribute to affordable agriculture by supporting cost-effective farming practices and encouraging crop diversification. However, the inherently recalcitrant nature of certain legume species poses a significant challenge, often resulting in limiting their overall production potential [27, 28]. Crops like chickpea, gram, and beans are being actively researched to improve resistance to various destructive diseases, including damping-off, collar rot, chlorotic dwarf virus, leaf roll virus, redleaf virus, and yellow mosaic virus, thereby strengthening crop resilience and ensuring better productivity. Recently, a marker-assisted backcrossing approach was employed to introgress the *loc4* and *loc1* loci into chickpea breeding programs for the development of new cultivars resistant to fusarium wilt and ascochyta blight [29, 30]. *VrTAF5* and *VrPGIP2* genes were introgressed for mungbean crop improvement targeting resistance to Cercospora leaf spot and bruchids, using a combination of gene pyramiding and marker-assisted backcrossing approaches [31]. In common bean, *I* and *bc-12* genes were transferred to confer resistance to mosaic virus and anthracnose [32]. Virus-resistant blackgram varieties were developed through marker-assisted backcrossing, and associated QTLs were identified [33]. Meta-QTLs were detected for partial resistance to white mold in common bean [34]. Genome-Wide Association Study (GWAS) was employed for single nucleotide polymorphism (SNP) identification to confer resistance in soybean against bacterial, viral, fungal diseases and nematode infestations [35]. Gene identification for frogeye leaf spot in soybean was conducted using a Genome-Wide Association Study (GWAS) [36]. Insect-resistant chickpea cultivars were developed by introgressing the *cry1Ac* gene using a marker-assisted backcrossing approach [37].

2.3 Resistance to biotic stress in oilseed crops

A diverse range of oilseed crops play a crucial role in providing edible oils for human consumption. Commonly cultivated oilseeds, such as soybean, groundnut, sunflower, mustard, sesame, safflower, and rice bran, serve as key sources of dietary fats. Beyond their dietary benefits, many oilseed crops hold significant industrial importance, being used in the production of biofuels, valuable metabolites, and diverse fatty acid profiles for various commercial applications. Oilseeds and their by-products majorly oils are the most valuable commodity in today's growing population. However, the industry faces a major challenge due to the decline in oil production from these crops. As a result, recent efforts in oilseed crop improvement have concentrated on enhancing seed quality, oil content, and oil composition [38]. Meanwhile, various QTLs and genes, such as *arahy.AF6N0J, arahy.XA63TZ*, etc., were identified against leaf spot in groundnut [39]. To predict the *Rcs3* gene responsible for conferring resistance in soybean cultivars, several SNPs were identified using a GWAS approach [40]. Molecular markers were employed to map and identify novel genes in soybean breeding programs for resistance to frogeye leaf spot [41]. Association mapping in mustard was employed to build resistance against stem rot [42]. GWAS analysis in sunflower identified marker-trait associations linked to Sclerotinia head rot, Phomopsis stem canker, and Phomopsis head rot [43].

2.4 Resistance to biotic stress in horticultural crops

Horticultural crops are generally classified into three main groups: fruits, vegetables, and ornamental or flowering plants. Among all crop types, horticultural crops especially fruits and vegetables are recognized for their rich nutritional profiles, containing essential dietary fiber, vitamins, and health-promoting phytochemicals [44]. This nutritional density often places them in the category of superfoods. However, the natural biochemical and physiological adaptations of plants are often insufficient to fully protect them against biological stresses. Therefore, it is essential to deepen our understanding of how horticultural crops respond to these challenges and to identify cultivars with enhanced tolerance. Initially, the investigation of resistant genes for late bight in potato was studied and *avl1* + *chc1* genes were found to confer resistance in the hybrids studied [45]. Through marker-assisted backcross breeding approach, *Ty-2* gene was introduced into tomato to confer leaf curl disease [46]. QTLs associated with viral diseases in cucumber were studied by using Genome-Wide Association Studies [47]. Similarly, *Fom-2* gene was introgressed in a susceptible cultivar through marker-assisted backcrossing approach to confer fungal diseases in melons [48]. A method of gene pyramiding was used in rose against fungal black spot races and identified resistance locus *Rdr3* [49]. Genes conferring resistance to Citrus tristeza virus were introduced into citrus, and four markers associated with the disease were utilized in a marker-assisted breeding approach [50]. A gene pyramiding approach of *Ppa3 and Xca1bo genes* was implemented to confer resistance to Downy mildew and Black rot in cauliflower [51]. A study identified the *Rpi2* gene for late blight resistance in potato using a molecular breeding approach [52]. A study identified several QTLs in banana associated with resistance to black leaf streak disease, which will be further utilized in marker-assisted selection [53]. Genome-Wide Association Studies (GWAS) in *Capsicum* resulted in the identification of several QTLs associated with various traits, including four specifically related to anthracnose resistance [54].

2.5 Resistance to biotic stress in fiber crops

Fiber crops like cotton, jute, and sunn hemp play a vital role in the agricultural sector, with fiber quality being a key target in breeding programs due to its direct impact on the national economy. In jute crop, traditional breeding approaches have been employed but the success rate was low as compared to other crops due to its complex genome and recalcitrant nature [55]. Meanwhile, in cotton, several improved cultivars have been released. Some researchers continue to investigate advanced breeding approaches to tackle the serious disease challenges that cause significant yield losses in cotton production. Nine QTLs associated with bacterial blight in cotton were identified in germplasm that provide a clear evidence that Asiatic cotton exhibits resistance to *Xanthomonas citri* pv. *malvacearum* (Xcm) race 18 [56]. Similarly, multiple stable QTLs were associated with two wilt diseases in cotton [57] by employing Genome-Wide Association Studies. Previous reviews have indicated that the presence of a single dominant gene conferring wilt resistance in cotton is often sufficient to significantly reduce infestation levels. Notably, *CG02* has been identified as a candidate gene associated with wilt resistance in cotton [58, 59]. Breeding programs were carried out to improve sunn hemp variety against major pests and diseases [60]. Furthermore, a study revealed 71 QTLs associated with root knot nematode in cotton by an approach of meta-analysis [61]. Future research aimed at developing pest- and pathogen-resistant cotton varieties remains essential, as the mapping of resistance loci within the cotton genome is still limited. Advancing molecular breeding approaches will be crucial to address these gaps [62].

3. Conclusion and future prospects

Advancements in breeding techniques have greatly strengthened crop protection in key agricultural crops, leading to improvements in food security, health, nutrition, and economic stability. This chapter offers a comprehensive review of previous research on crop improvement, with a particular focus on methods used to develop climate-resilient varieties. Although these accomplishments are notable, challenges persist, particularly the rapid evolution of pathogens and the ongoing need for broad-spectrum, durable resistance. To overcome these obstacles, future breeding approaches must integrate cutting-edge technologies, especially multi-omics breeding. This approach, integrating genomics, transcriptomics, proteomics, metabolomics, and phenomics, will allow breeders to gain a comprehensive understanding of how crops respond to environmental stresses at both the molecular and physiological levels. For instance, linking genomic data with phenotypic traits will enable the identification of genes responsible for both biotic resistance and stress tolerance, facilitating the development of crops that are resistant to pests and adaptable to changing climatic conditions. Moreover, high-throughput genotyping and phenotyping will remain essential for accelerating the identification and selection of resistant lines. Technologies such as drones and advanced imaging systems will enable real-time measurement of crop traits, speeding up data collection and improving precision, thereby supporting a more accurate genomic selection and accelerating the development of superior varieties. By adopting these strategies, future crop breeding efforts will ensure the creation of resilient crops capable of meeting the escalating food demands of a growing global population in the coming decades.

Author details

Vividha M. Sirsat[1*], Aditi R. Rajane[2], Pallavi R. Sasane[3] and Bhagyashree P. Bahatkar[4]

1 Biotechnology Center, Department of Agricultural Botany, Dr. Panjabrao Deshmukh Krishi Vidyapeeth, Akola, Maharashtra, India

2 P. R. Pote, College of Agriculture, Amravati, Maharashtra, India

3 Government College of Agriculture, Nandurbar, Maharashtra, India

4 SKR Agrotech, Wardha, Maharashtra, India

*Address all correspondence to: vividhasirsat@gmail.com

References

[1] Nature Communications. Feeding the future global population. Nature Communications. 2024;**15**:222. DOI: 10.1038/s41467-023-44588-y

[2] Nejat N, Mantri N. Plant immune system: Crosstalk between responses to biotic and abiotic stresses the missing link in understanding plant defense. Current Issues in Molecular Biology. 2017;**23**:1-16

[3] Das G, Rao GJN. Molecular marker assisted gene stacking for biotic and abiotic stress resistance genes in an elite rice cultivar. Frontiers of Plant Science. 2015;**6**:698

[4] Collard BCY, Mackill DL. Marker-assisted selection: An approach for precision plant breeding in the twenty-first century. Philosophical Transactions of the Royal Society B: Biological Sciences. 2008;**363**(1491):557-572

[5] Smith J, Johnson A, Brown P. DNA-based markers in plant breeding: Applications and advancements. Journal of Molecular Plant Breeding. 2020;**45**(2):123-135

[6] Patranik D, Khuruana P. Wheat biotechnology: A minireview. EJB Electronic Journal of Biotechnology. 2001;**4**(2):1-4

[7] Figueroa M, Hammond-Kosack KE, Solomon PS. A review of wheat diseases-a field perspective. Molecular Plant Pathology. 2018;**19**(6):1523-1536. DOI: 10.1111/mpp.12618. Epub 2017 Dec 26

[8] McIntosh RA, Dubcovsky J, Rogers JW, Morris C, Xia CX. Catalogue of Gene Symbols for Wheat: 2017 Supplement. 2017

[9] Yadawad A, Gadpale A, Hanchinal RR, Nadaf HL, Desai SA, Biradar S, et al. Pyramiding of leaf rust resistance genes in bread wheat variety DWR 162 through marker assisted backcrossing. Indian Journal of Genetics and Plant Breeding. 2017;**77**(02):251-257

[10] Bhawar KB, Vinod SJB, Singh AK, Sivasamy S, Singh M, Prabhu KV, et al. Molecular marker assisted pyramiding of leaf rust resistance genes *Lr19* and *Lr28* in wheat bread (*Triticum aestivum* L.) variety HD2687. Indian Journal of Genetics. 2011;**71**(4):304-3011

[11] Ali MA, Shahzadi M, Zahoor A, Dababat AA, Toktay H, Bakhsh A, et al. Resistance to cereal cyst nematodes in wheat and barley: An emphasis on classical and modern approaches. International Journal of Molecular Sciences. 2019;**20**(2):432

[12] Shahin AA, Omara RI, Omar HA, et al. Evaluation of effectiveness resistance genes in wheat genotypes using marker-assisted selection for stripe rust resistance breeding. BMC Plant Biology. 2024;**24**:785

[13] Qian Z, Han G, Ningning Y, Liu C, Han R, Jameson PE, et al. Fine mapping of the powdery mildew resistance gene PmXQ-0508 in bread wheat. The Crop Journal. 2024;**12**(4):1176-1184. ISSN 2214-5141

[14] Ji Z-j, Yang S-d, Zeng Y-x, Liang Y, Yang C-d, Qian Q. Pyramiding blast, bacterial blight and brown planthopper resistance genes in rice restorer lines. Journal of Integrative Agriculture. 2016;**15**(7):1432-1440

[15] Suvendhu DS, Divya D, Durga Rani CV, Dayakar Reddy T,

Visalakshmi V, Cheralu C, et al. Characterization of gall midge resistant rice genotypes using resistance gene specific markers. Journal of Experimental Biology and Agricultural Sciences. 2014;**2**(4):439-446

[16] Harathi PN, Raja Sekhar P, Hari Prasad KV, Bhaskar Reddy BV, Vineetha U, Ravindra Reddy B. Gall midge resistance gene identification in promising rice genotypes using gene specific marker. Biological Forum–An International Journal (SI-AAEBSSD-2021). 2021;**13**(3b):300-303

[17] Chen Q, Zeng G, Hao M, et al. Improvement of rice blast and brown planthopper resistance of PTGMS line C815S in two-line hybrid rice through marker-assisted selection. Molecular Breeding. 2020;**40**:21

[18] Kumar S, Kumar A. Advances in genomic tools for plant breeding: Harnessing DNA technologies for crop improvement. Biological Research. 2024;**57**(1):62

[19] Zhao X, Tan G, Xing Y, et al. Marker-assisted introgression of *qHSR1* to improve maize resistance to head smut. Molecular Breeding. 2012;**30**:1077-1088. DOI: 10.1007/s11032-011-9694-3

[20] Xu Z, Hua J, Wang F, Cheng Z, Meng Q, Chen Y, et al. Marker-assisted selection of *qMrdd8* to improve maize resistance to rough dwarf disease. Breeding Science. 2020;**70**(2):183-192. DOI: 10.1270/jsbbs.19110. Epub 2020 Feb 26

[21] Gorthy S, Narasu L, Gaddameedi A, Shrama HC, Kotla A, Deshpande SP, et al. Introgression of shoot Fly (*Alherigona soccata* L. Moench) resistance QTLs into elite post-rainy season sorghum varieties using marker assisted backcrossing (MABC). Frontiers in Plant Science. 2017;**8**:1494

[22] Li J, Zhang L, Guo X, Zhang J, Wang S, Sun X, et al. Identification of novel QTL contributing to resistance against Aspergillus flavus in maize (*Zea mays* L.) using an enlarged genotype panel. Journal of Integrative Agriculture. ISSN 2095-3119. 2025. DOI: 10.1016/j.jia.2025.01.002

[23] Sun H, Zhai L, Teng F, Li Z, Zhang Z. qRgls1.06, a major QTL conferring resistance to gray leaf spot disease in maize. The Crop Journal. 2021;**9**(2):342-350. ISSN 2214-5141,. DOI: 10.1016/j.cj.2020.08.001

[24] Hautsalo J, Novakazi F, Jalli M, et al. Pyramiding of scald resistance genes in four spring barley MAGIC populations. Theoretical and Applied Genetics. 2021;**134**:3829-3843. DOI: 10.1007/s00122-021-03930-y

[25] Amezrou R, Verma RPS, Chao S, et al. Genome-wide association studies of net form of net blotch resistance at seedling and adult plant stages in spring barley collection. Molecular Breeding. 2018;**38**:58. DOI: 10.1007/s11032-018-0813-2

[26] Li G, Xiangyang X, Tan C, Carver BF, Bai G, Xuewen Wang J, et al. Identification of powdery mildew resistance loci in wheat by integrating genome-wide association study (GWAS) and linkage mapping. The Crop Journal. 2019;**7**(3):294-306

[27] Nivya VM, Shah JM. Recalcitrance to transformation, a hindrance for genome editing of legumes. Frontiers in Genome Editing. 2023;**5**:1247815. DOI: 10.3389/fgeed.2023.1247815

[28] Semba RD, Ramsing R, Rahman N, Kraemer K, Bloem MW. Legumes

as a sustainable source of protein in human diets. Global Food Security. 2021;**28**:100520

[29] Mannur DM, Babbar A, Thudi M, Sabbavarapu MM, Roorkiwal M, Yeri SB, et al. Super Annigeri 1 and improved JG 74: Two fusarium wilt-resistant introgression lines developed using marker-assisted backcrossing approach in chickpea (*Cicer arietinum* L.). Molecular Breeding. 2019;**39**:2. DOI: 10.1007/s11032-018-0908-9

[30] Varshney RK, Mohan SM, Gaur PM, Chamarthi SK, Singh VK, Srinivasan S, et al. Marker-assisted backcrossing to introgress resistance to fusarium wilt race 1 and ascochyta blight in C 214, an elite cultivar of chickpea. The Plant Genome. 2014;**7**(1):1-11

[31] Laosatit K, Yimram T, Keawwongwal A, Srichan M, Amkul K, Tanadul O-u-m, et al. Development of pyramided mung bean lines carrying resistance genes for cercospora leaf spot disease and bruchids. Chilean Journal of Agricultural Research. 2024;**84**(5):644-652

[32] Hegay S. Diversity of beans grown in kyrgyzstan and marker-aided breeding for resistance to bean common mosaic virus and anthracnose [thesis]. Faculty of Landscape Planning, Horticulture and Agricultural Science Department of Plant Breeding; 2013

[33] Subramaniyan R, Narayana M. Marker- assisted backcross breeding to introgress bruchid [*Callosobruchus maculatus* (F.)] and mungbean yellow mosaic virus (MYMV) disease resistance in blackgram [*Vigna mungo* (L.) Hepper] variety MDU 1. Australasian Plant Pathology. 2025;**54**:63-72. DOI: 10.1007/s13313-025-01018-5

[34] Vasconcellos RCC, Oraguzie OB, Soler A, Arkwazee H, Myers JR, Ferreira JJ, et al. Meta-QTL for resistance to white mold in common bean. PLoS One. 2017;**12**(2):e0171685. DOI: 10.1371/journal.pone.0171685

[35] Akram M, Kamaal N, Pratap A, Singh NP. Resistance status of mungbean [*Vigna radiata* (L.) Wilczek] advanced breeding materials against mungbean yellow mosaic India virus. Archives of Phytopathology and Plant Protection. 2021;**54**(19-20):2533-2546

[36] Sun M, Na C, Jing Y, Cui Z, Li N, Zhan Y, et al. Genome-wide association analysis and gene mining of resistance to China race 1 of frogeye leaf spot in soybean. Frontiers in Plant Science. 2022;**13**:867713

[37] Kaur A, Shrama U, Singh R, Vikal Y, Singh S, Malik P, et al. Introgressing Cry1Ac for pod borer resistance in chickpea through marker-assisted backcross breeding. Frontiers in Genetics. 2022;**13**:847647. DOI: 10.3398/fgene.2022.847647

[38] Gupta SK, editor. Breeding Oilseed Crops for Sustainable Production Opportunities and Constraints. Elsevier: Academic Press; 2016

[39] Zhang H, Chu Y, Dang P, et al. Identification of QTLs for resistance to leaf spots in cultivated peanut (*Arachis hypogaea* L.) through GWAS analysis. Theoretical and Applied Genetics. 2020;**133**:2051-2061

[40] Mian MAR, Bond J, Joobeur T, Mengistu A, Wiebold W, Shannon G, et al. Identification of soybean genotypes resistant to *Cercospora sojina* by field screening and molecular markers. Plant Disease. 2009;**93**:408-411

[41] Yang W, Weaver DB, Nielsen BL, Qiu J, Salamini F. Molecular mapping of a new gene for resistance to frogeye

leaf spot of soya bean in 'Peking'. Plant Breeding. 2001;**120**(1):73-78

[42] Rana K, Atri C, Gupta M, Akhatar J, Sandhu PS, Kumar N, et al. Mapping resistance responses to Sclerotinia infestation in introgression lines of *Brassica juncea* carrying genomic segments from wild *Brassicaceae B. Fruticulosa*. Scientific Reports. 2017;7:5904. DOI: 10.1038/s41598-017-05992

[43] Filippi CV, Corro Molas A, Dominguez M, Colombo D, Heinz N, Troglia C, et al. Genome-wide association studies in sunflower: Towards sclerotinia sclerotiorum and Diaporthe/Phomopsis resistance breeding. Genes. 2022;**13**(12):2357

[44] Shivran M, Shivran U, Singh N. Breeding approaches in fruit crops improvement. The Pharma Innovation Journal. 2022;**11**(4):2034-2038

[45] Su Y, Viquez-Zamora M, den Uil D, et al. Introgression of genes for resistance against *Phytophthora infestans* in diploid potato. American Journal of Potato Research. 2020;**97**:33-42. DOI: 10.1007/s12230-019-09741-8

[46] Sowjanya BA, Sridevi O. Introgression of tomato leaf curl virus (ToLCV) resistant gene into two cultivated tomato (*Solanum lycopersicum* L.) varieties through marker assisted backcross breeding. Biological Forum–An International Journal. 2021;**13**(1):211-226

[47] Monnot S, Cantet M, Mary-Huard T, Moreau L, Lowdon R, Van Haesendonck M, et al. Unravelling cucumber resistance to several viruses via genome-wide association studies highlighted resistance hotspots and new QTLs. Horticulture Research. 2022;**9**:uhac184

[48] Sousaraei N, Ramshini H, Lotfi M, et al. Marker assisted backcrossing for introgression of fusarium wilt resistance gene into melon. Euphytica. 2018;**214**:7. DOI: 10.1007/s10681-017-2080-4

[49] Whitaker VM, Bradeen JM, Debener T, et al. *Rdr3*, a novel locus conferring black spot disease resistance in tetraploid rose: Genetic analysis, LRR profiling, and SCAR marker development. Theoretical and Applied Genetics. 2010;**120**:573-585. DOI: 10.1007/s00122-009-1177-0

[50] Ohta S, Endo T, Shimada T, Fujii H, Shimizu T, Kuniga T, et al. PCR primers for marker assisted backcrossing to introduce a CTV resistance gene from *Poncirus trifoliata* (L.) Raf. into *citrus*. Journal of Japanese Society for Horticultural. 2011;**80**(3):295-307

[51] Saha P, Ghoshal C, Saha ND, Verma A, Srivastava M, Kalia P, et al. Marker-assisted pyramiding of downy mildew-resistant gene *Ppa3* and black rot-resistant gene *Xca1bo* in popular early cauliflower variety Pusa Meghna. Frontiers in Plant Science. 2021;**12**:603600

[52] Yang L, Wang D, Xu Y, Zhao H, Wang L, Cao X, et al. A new resistance gene against potato late blight originating from *Solanum pinnatisectum* located on potato chromosome 7. Frontiers in Plant Science. 2017;**8**:1729

[53] Carreel F, Martin G, Ravel S, Roussel V, Pages C, Habas R, et al. Identification of resistance QTLs to black leaf streak disease (due to *Pseudocercospora fijiensis*) in diploid bananas (*Musa acuminata*). Horticulturae. 2024;**10**(6):608

[54] Jo J, Kim GW, Back S, et al. Exploring horticultural traits and disease resistance

in *Capsicum baccatum* through segmental introgression lines. Theoretical and Applied Genetics. 2023;**136**:233. DOI: 10.1007/s00122-023-04422-x

[55] Author Dempsey JM. Fibre Crops. Gainesville, FL: University Presses of Florida; 1975. p. 457

[56] Abdelraheem A, Zhu Y, Zeng L, et al. Identification of new genetic sources of resistance to bacterial blight race 18 in diploid Asiatic cotton and resistance transfer to tetraploid upland cotton (*Gossypium hirsutum*). Euphytica. 2024;**220**:85

[57] Abdelraheem A, Elassbli H, Zhu Y, et al. A genome-wide association study uncovers consistent quantitative trait loci for resistance to verticillium wilt and fusarium wilt race 4 in the US upland cotton. Theoretical and Applied Genetics. 2020;**133**:563-577

[58] Li T, Ma X, Li N, Zhou L, Zheng L, Han H, et al. Genome-wide association study discovered candidate genes of verticillium wilt resistance in upland cotton (*Gossypium hirsutum* L.). Plant Biotechnology Journal. 2017;**15**:1520-1532

[59] Zhang J, Fang H, Zhou H, et al. Genetics, breeding, and marker-assisted selection for verticillium wilt resistance in cotton. Crop Science. 2014;**54**(4):1289-1303

[60] Sanjoy S, Mitra J, Pandey SK. JRJ 610 (Prankur): A new sunnhemp (*Crotalaria juncea* L.) variety for high yielding and superior quality fibre. Journal of Pharmacognosy and Phytochemistry. 2018;7(4):2354-2357

[61] Said JI, Lin Z, Zhang X, Song M, Zhang J. A comprehensive meta QTL analysis for fiber quality, yield, yield related and morphological traits, drought tolerance, and disease resistance in tetraploid cotton. BMC Genomics. 2013;**14**:776. DOI: 10.1186/1471-2164-14-776

[62] Kumar P, He Y, Singh R, et al. Fine mapping and identification of candidate genes for a QTL affecting meloidogyne incognita reproduction in upland cotton. BMC Genomics. 2016;**17**:567. DOI: 10.1186/s12864-016-2954-1

Section 3

Genetic Engineering and Novel Approaches

Chapter 5

Genetic Engineering of Agricultural Crops for Food and Nutritional Security

Muthukumar Elankumaran, Anbuselvan Antony, Shivaanie Nandakumar, Vidyasagar Pari, Prosenjeet Chakraborthy, Sundarasamy Dhanapal, Chinnappan Deepa Joan of Arc and Krishnagowdu Saravanan

Abstract

Genetic engineering has been established as a believable technology in modern agriculture, permitting the development of nutritional superiority in various crops. Traditional breeding methodologies have long been used to increase crop traits, but they are often inefficient and incomplete due to genetic compatibility. In contrast, genetic engineering permits accurate variations to plant genomes, helping enrich essential nutrients such as vitamins, minerals, amino acids, and innumerable bioactive constituents. This technology addresses widespread malnutrition and micronutrient deficiencies, mainly in the developing countries. Distinguished successes in genetic engineering-mediated biofortification include the development of Golden Rice, engineered to produce β-carotene, high-lysine maize, zinc, and iron-fortified wheat. Recent advancements in CRISPR/Cas9 gene editing and transgenic platforms have significantly broadened the scope for precise nutritional enhancement in crop species. Improved production of essential fatty acids, antioxidants, and secondary metabolites in genetically modified crops has shown significant potential in promoting human health and preventing malnutrition-related disorders. In addition to contributing to nutritional security, genetic engineering has increased crop output, stress tolerance, and resistance to pests and diseases. This confirms increased food and nutritional security in the face of improving population and environmental challenges. However, despite this auspicious enhancement, regulatory issues and ethical and public concerns pose barriers to the mainstream implementation of genetically modified crops. This chapter examines the most recent advances in genetic engineering for nutritional enhancement in crops, highlighting key methods, successful case studies, and forthcoming scenarios.

Keywords: genetic engineering, biofortification, genetically modified crops, CRISPR/Cas9, food and nutritional security

1. Introduction

Food and nutritional deficiencies continue to be a substantial global health concern, with over 2 billion people suffering from micronutrient malnutrition, particularly in developing countries [1]. Staple crops, though calorie-rich, often lack essential nutrients like vitamin A, iron, and zinc. Enhancing the nutritional content of crops is, therefore, a strategic solution to address food and nutritional security. Current developments in plant biotechnology have placed genetic engineering as an auspicious method for nutritional enhancement in crops, permitting the precise addition or improvement of essential mineral nutrients [1, 2]. Traditional breeding has brought good results in increased yield and some quality traits, but actual methods are limited by genetic compatibility, long polygenic inheritance, and long breeding cycles [3–6]. Such limitations especially slow down the improvement of micronutrient characteristics when those characteristics do not exist within the gene pool of the target crops [7]. In addition, traditional methods are often inefficient in stacking multiple desirable traits into a single variety, enhancing nutrient levels, tolerating abiotic stress, and resisting pests [8]. Genetic engineering can eliminate all constraints of traditional breeding and allow direct, targeted manipulation in plant genomes. Precision breeding has been further modernized by CRISPR/Cas9, which allows site-specific gene modification without inserting foreign DNA [9]. These developments thus provide a scalable route for improving crop nutritional quality, enhancing human health, and ensuring sustainable agriculture [10]. Genetic engineering opens new creativity toward solving pressing global challenges: food security, malnutrition, climate change, and sustainable agriculture through direct modification of living organisms' genes [11]. This chapter will review in detail the major methods of genetic engineering: transgenic technology, CRISPR-Cas9 genome editing, RNA interference, and synthetic biology. The chapter also gives a perspective on how these technologies can be applied to yield enhancement, stress tolerance, biofortification, and the future scenario of food systems worldwide.

2. Genetic engineering tools and techniques

2.1 Transgenic technology

Transgenic technology includes permanently incorporating a foreign gene into the DNA of a host organism. The nucleotide sequence of this gene may come from any other species or even an entirely different kingdom (for example, bacterial genes can be inserted into plants) and is typically used to bestow new characteristics such as improved nutritional content, resistance to pests, or tolerance to abiotic stresses on an already existing plant trait [12]. The two most common delivery methods are *Agrobacterium*-mediated transformation, which exploits the natural phenomenon of transfer by *Agrobacterium tumefaciens*, wherein one part of its Ti plasmid gets integrated into the cells of plants. This method nearly always applies to dicot plants and some monocots with minor changes [13]. Biolistic or gene gun method: This method physically shoots gold- or tungsten-coated DNA particles into plant tissues, mostly applicable for monocots such as maize and rice [14]. Combining scientific innovation with policy and public communication is crucial because, despite its effectiveness, transgenic technology has been criticized and is subject to regulatory obstacles because of worries about ecological safety and food labeling [15].

2.2 CRISPR/Cas9 genome editing

CRISPR (Clustered Regularly Interspaced Short Palindromic Repeats)/Cas9 technology is a pioneering technique for precise and current genome editing. In contrast to transgenic techniques, CRISPR allows for altering inherent plant genes without principally presenting foreign DNA, making it more suitable under current governing and regulatory frameworks in many nations [16, 17]. A guide RNA (gRNA) is designed and synthesized for efficient gene knock-in or knockout technology in the plant genome of our desired target. The gRNA and Cas9 protein complex create a double-strand break (DSB) at the desired target site in the genome. After the break is repaired by either homology-directed repair (HDR) or nonhomologous end joining (NHEJ) process [17]. CRISPR/Cas9 has been working to enhance iron content in rice by editing the *OsVIT2* [18]. Increase oleic acid levels in soybean by knockout of *FAD2-1A/B* [19]. Decrease anti-nutritional factors like phytic acid in cereals [20]. Its ease of use, instantaneous editing of multiple genes, and decreased off-target effects have made CRISPR a crucial tool in food and nutritional security [21].

2.3 RNA interference (RNAi)

RNA interference (RNAi) is a gene silencing technique that silences specific genes' expression after transcription. To achieve this, double-stranded RNA (dsRNA) is introduced and processed into small interfering RNAs (siRNAs). These siRNAs then bind to complementary mRNAs and trigger their degradation [22]. RNAi has been used extensively in biofortification, for it reduces anti-nutritional compounds such as phytic acid within rice and maize (*IPK1* silencing) for mineral absorption improvement [23]. Competitive metabolic pathways, such as anthocyanins and flavonoids, are silenced, increasing secondary metabolite content. Genes causing allergens or toxic metabolites can be silenced (e.g., *delta-cadinene synthase* in cotton to reduce gossypol) [24]. RNAi is functional because it targets genes and does not introduce permanent changes to the genome [25].

2.4 Synthetic biology and metabolic engineering

Synthetic biology combines engineering with molecular biology to fashion new biological systems [26]. It is metabolic engineering, as a subfield, that manipulates cellular pathways to improve or redirect cells for the production of desired compounds like vitamins, amino acids, and antioxidants [27]. This approach has been instrumental in designing new biosynthetic pathways for anthocyanins in rice and tomatoes [28]. Reconstructed vitamin B9 (folate) pathways in staple crops for congenital disabilities and anemia [29]. Engineering oilseed crops like *Camelina sativa* to produce long-chain omega-3 fatty acids [30]. Pathway compartmentalization strategies, promoter libraries, and gene circuits should allow synthetic biology to tailor plant nutrition [31].

3. Biofortification through genetic engineering

Biofortification refers to enhancing crops' nutritional value through classical breeding or modern genetic engineering approaches [32]. Biofortification aims to increase the accumulation of essential vitamins, minerals, and other nutrients in crops, making them more nourishing for human consumption and maintaining nutritional security [33]. Genetic engineering allows for the introduction of specific genes or transcription factors (TFs) that enhance nutrient content [34], particularly in staple crops, pulses or legumes, and horticultural crops that form the basis of the human diet, such as rice, wheat, maize, soybean, peanut, banana, and tomato [35]. This strategy has gained significant attention for addressing malnutrition, especially in regions where people lack access to foods rich in essential micronutrients [36]. Biofortification varies from conventional fortification in that it aims to improve the nutrient content of crops directly in the field rather than through postharvest processes like adding supplements to food products [37, 38]. Significant biofortification objectives include:

- *Decrease in micronutrient deficiency-related problems*: Aiming at the development of vital micronutrients and secondary metabolites such as vitamin A, iron, zinc, folate, anthocyanins, carotenoids, flavonoids, and polyphenolic contents [39].

- *Sustainability*: Enlightening the nutrient content of crops in ways that do not require the constant adding of peripheral additions [40].

- *Upgraded crop productivity*: It simultaneously enhances crop yields and improves nutritional value, thereby contributing to both nutritional security and the reduction of malnutrition [41].

- *Improved nutrient bioavailability*: Confirming that the added nutrients are in a bioavailable form, meaning they can be effectively absorbed and used by the human body [42]. The biofortification of crops through genetic engineering technology was summarized in **Table 1**.

3.1 Golden Rice: A case study in vitamin A enrichment

Golden Rice was established to eradicate vitamin A deficiency, a critical health problem, mainly in Southeast Asia and sub-Saharan Africa. β-Carotene is naturally present in the chloroplasts of plant leaves but is not usually present in important amounts in rice grains. Through cloning and transformation of these genes, Golden Rice can produce β-carotene in the endosperm of rice, encompassing the significant nutritional value of vitamin A. Golden Rice has experienced extensive trials and regulatory assessments in numerous nations and has verified the potential to eradicate the frequency of vitamin A deficiency, which can cause blindness and reduce the normal immune response in the human body [43].

3.2 High-lysine maize and quality protein maize

Maize (*Zea mays*) is a staple food worldwide [62], but the quality and quantity are very low, particularly in tryptophan and lysine, essential amino acids for the human diet. To avoid this issue, quality protein maize (QPM) addresses this deficiency [63]. The initial discovery was the *opaque-2* (*o2*) mutant, which enhances the lysine and tryptophan content by decreasing the α-zein proteins and upregulating non-zein proteins in maize. But, *o2* mutants have soft, chalky kernels [64]. The modifiers were introduced through classical breeding and marker-assisted selection processes to restore kernel rigidity while retaining high nutritional quality [65]. Genetic

Crop(s)	Target trait	Genes/transcription factor (TF)	Function/activity	References
Rice (*Oryza sativa* (L.))	Provitamin A (β-carotene)	*psy*, *crtI*	Enhances carotenoid biosynthesis in rice endosperm	[43]
	Iron accumulation	*Ferritin*, *OsNAS1*, *OsYSL2*	Enhances iron storage and transport	[44, 45]
	Zinc accumulation	*OsNAS2*	Increases zinc chelation and transport	[45]
	Folate (vitamin B9)	*GTPCHI*, *ADCS*	Enhances folate biosynthesis	[46]
	Tryptophan	Feedback-insensitive *anthranilate synthase*	Increases tryptophan content	[47]
	Iron and zinc	*Barley nicotianamine synthase* gene *HvNAS1*	Increases iron and zinc concentrations	[48]
Soybean	Tryptophan	Feedback-insensitive *anthranilate synthase*	Enhances amino acid biosynthesis	[49]
	Vitamin E	*γ-TMT*	Enhances α-tocopherol content	[50–52]
	Flavonoids	*GmMYB176*	Regulates isoflavonoid biosynthesis	[53]
	Oil content	*DGAT1A*	Increases triacylglycerol biosynthesis	[54]
	Omega-3 fatty acids	*Fatty acid desaturase* genes (FAD3, FAD2)	Increases ALA and EPA content	[55]
Potato	Provitamin A	*crtB*, *crtI*, *crtY*	Stimulates carotenoid biosynthesis	[56]
Cassava	Provitamin A	*crtB*	Elevates beta-carotene in tubers	[57]
Tomato	Flavonols	*AtMYB12*	Activates flavonol pathway	[58]
	Carotenoids	*LCYb*, *PSY1*	Enhances lycopene and beta-carotene content	[59]
Banana	Provitamin A	*MaPSY1*, *MaLCYB*	Increases carotenoid levels	[60]
Kale	Flavonoids	*AtMYB12*	High flavonol accumulation	[61]

Table 1.
Biofortification of crops using genetic engineering technology.

engineering has complemented QPM breeding. For example, overexpression of *ferritin* or downregulation of zein genes using RNAi techniques has enhanced lysine accumulation [66].

3.3 Iron- and zinc-enriched wheat and rice

Iron and zinc are essential micronutrients that are frequently deficient in the diets of people in developing countries. To fight these deficiencies, researchers have established genetically improved crops such as iron- and zinc-enriched wheat and rice [54]. The iron-enriched rice was established by introducing genes involved in the biosynthesis and storage of ferritin [67]. By enhancing ferritin expression, researchers increased the iron content in rice grains, making it a potential solution for iron-deficiency anemia, which is a widespread issue, particularly among women and children in developing regions [68]. Likewise, zinc-enriched wheat has been established by introducing genes for zinc transporters, which improve the uptake and storage of zinc in the wheat [69, 70]. These genetically modified crops could offer a sustainable solution for alleviating zinc deficiency.

3.4 Enhancement of secondary metabolites and antioxidants

In addition to improving essential vitamins and minerals, genetic engineering is also being used to enhance the production and accumulation of bioactive metabolites (secondary metabolites) in crops [71]. These compounds include flavonoids, carotenoids, and phenolic acids, which have significant health benefits in humans, such as antioxidant, anticancer, anti-inflammatory, antimicrobial, antiviral, etc. [72]. For example, researchers have genetically modified tomatoes to produce higher levels of lycopene, a potent antioxidant associated with a reduced risk of cancer and heart-related disease [73]. Likewise, researchers have been working on biofortifying crops such as spinach and broccoli to enhance the levels of various polyphenolic compounds known for their anti-inflammatory and antioxidant effects [74]. Improving secondary metabolites through genetic engineering can enhance the health benefits of commonly consumed crops, providing further safety against various chronic diseases such as cancer, diabetes, and cardiovascular disease [75].

4. Impact on food and nutritional security

The universal food system has many encounters, including population growth, climate change, soil degradation, and socioeconomic disparities that result in persistent food insecurity and undernourishment [76]. Genetic engineering technology proposes groundbreaking solutions that improve crop productivity, nutritional content, and resilience to environmental stresses [77].

4.1 Addressing global malnutrition and hidden hunger

Malnutrition remains a persistent worldwide problem, with over 2 billion people suffering from micronutrient deficiencies—called "hidden hunger" [78]. These deficiencies, chiefly vitamin A, iron, zinc, and iodine, impair immune function, cognitive development, and overall health [79]. Genetic engineering delivers a maintainable and cost-effective solution by biofortifying staple crops to enhance their micronutrient content [80]. Field trials and human feeding studies have confirmed its efficacy and safety [81]. Another example is iron- and zinc-biofortified rice and wheat, developed by overexpressing ferritin genes [82] from legumes and *nicotianamine synthase* (*NAS*) [83], which improve mineral bioavailability.

4.2 Contribution to public health

Beyond their agricultural benefits, genetically modified crops can increase public health by mitigating nutrient deficiencies and chemical exposure and avoiding foodborne diseases [84]. The consumption of biofortified crops has been shown to reduce micronutrient deficiency-related illnesses, mainly in children and women [85]. These crops are combined into national nutrition strategies in several countries, including Bangladesh, the Philippines, and Nigeria [86]. Additionally, genetically engineered crops have helped decrease the occurrence of mycotoxins—carcinogenic mixtures produced by fungi—by introducing resistance to *Aspergillus flavus* in maize through RNAi or antifungal protein expression [87]. These reduce the incidence of liver cancer and other health issues associated with aflatoxin exposure. Health and environmental risk assessments of GM foods by major regulatory bodies, including the World Health Organization (WHO), Food and Agriculture Organization (FAO), and European Food Safety Authority (EFSA), have consistently affirmed the safety of permitted genetically modified crops when accurately evaluated [88]. Over 20 years of data show no long-established cases of adverse effects on human health from accepted GM crops [89]. In addition, reducing the ability to modify crops with allergenic proteins opens new chances to produce hypoallergenic foods [67]. The Golden Rice field trial was conducted in several locations in the Philippines and Bangladesh, demonstrating steady agronomic performance and higher provitamin A levels without productivity losses [48]. Similarly, HarvestPlus' iron- and zinc-rich wheat had 40–50% increased micronutrient content while maintaining yield parity with conventional types in South Asian climates [68]. These varieties have been released in India and Pakistan and are being used in government food programs [90]. With encouraging outcomes, larger implementation is hampered by regulatory delays, public mistrust, and the need for additional long-term nutrition and environmental impact assessments [91].

5. Challenges and concerns

Despite the research improvements and established benefits of genetically modified crops, their acceptance is repeatedly susceptible to several regulatory, socio-ethical, and ecological concerns [92]. The monitoring frameworks that are mainly governing genetically modified crops vary widely among nations, ranging from permissive to highly limiting [93]. Ethical and ecological concerns include unintentional gene flow to wild species, possible appearance of resistant pests or superweeds, and the perceived monopolization of agriculture by biotech firms, which may limit farmers' rights and agrobiodiversity [94]. Social debates also focus on food authority, reasonable access, and the long-term impacts of genetic modifications on ecological systems [95]. Public discernment and receipt remain major challenges, often predisposed by misrepresentation, cultural beliefs, and mistrust in institutions, rather than scientific evidence. Although frequent self-determining reviews—including those by the World Health Organization (WHO), Food and Agriculture Organization (FAO), and National Academies of Sciences (NAS)—confirm the safety of approved genetically modified crops for consumption and the environment, public skepticism continues, particularly in Europe and parts of Asia. Bridging this gap requires inclusive policymaking, clear risk communication, participatory stakeholder engagement, and consistent post-release monitoring to ensure that biotechnology advances align with social values and sustainability goals.

The precision of genome editing, particularly CRISPR/Cas systems, has significantly accelerated the development of nutritionally enhanced crop traits [96]. However, off-target mutations remain a concern, mainly due to unintended cleavage at genomic loci with sequence homology to guide RNAs. Recent high-fidelity CRISPR variants such as SpCas9-HF1, eSpCas9, HypaCas9, and base editors have substantially reduced these risks [97]. Complementary bioinformatic tools like CRISPR, GUIDE-seq, CIRCLE-seq, Cas-OFFinder, and CRISPR-P 2.0 are routinely employed to predict and validate off-target effects in plant genomes [98, 99]. Trait stacking for multi-nutrient biofortification poses additional challenges, including transgene silencing, promoter interference, metabolic crosstalk, and nutrient loading imbalances [100]. These complexities often hinder stable trait expression and long-term performance. Advanced solutions such as synthetic promoters, orthogonal gene circuits, independent regulatory elements, and CRISPR/Cas-mediated multiplex editing or homology-directed repair are being developed to overcome these obstacles [101]. For instance, CRISPR/Cas9 targeting of *OsNRAMP5* in rice has enhanced zinc accumulation while reducing cadmium uptake [102]. Ultimately, understanding and addressing these molecular bottlenecks is crucial for ensuring the safety, stability, and efficacy of nutritionally improved genetically engineered crops in diverse agricultural environments [92].

While genetic engineering offers powerful tools for improving crop nutrition, several critical challenges hinder its widespread acceptance and deployment. Public acceptance remains low in many regions due to misinformation, ethical concerns, and lack of trust in regulatory agencies, despite scientific consensus on Genetically Modified Organism (GMO) safety [103, 104]. Regulatory hurdles are significant, especially in the EU and parts of Asia, where gene-edited crops are subjected to lengthy approval processes similar to transgenics, delaying innovation [105]. Ecological risks, such as unintended gene flow to wild species, off-target effects, and biodiversity disruption, require rigorous long-term assessments [106]. Additionally, the economic concentration of agri-biotech patents among a few global firms creates intellectual property barriers, limiting access for smallholder farmers and

Feature	Traditional genetic engineering	CRISPR-mediated genome editing
Mechanism	Transgene insertion from other species	Precise editing of native DNA sequences
Foreign DNA	Often involves foreign genes (transgenic)	May involve no foreign DNA (cisgenic or non-transgenic)
Target specificity	Less predictable, may insert randomly	Highly specific and programmable targeting
Regulatory status	Heavily regulated in most countries	Often lighter regulation (e.g., USA, Japan)
Time to market	Longer due to strict biosafety assessments	Faster due to fewer regulatory hurdles
Consumer/public acceptance	Controversial; labeled in many countries	Generally higher acceptance if no foreign DNA involved
Use cases in nutrition	Golden Rice, high-lysine maize, iron-fortified wheat	CRISPR-edited tomatoes with GABA, enhanced flavonoids
Limitations	Risk of gene silencing, public mistrust, high cost	Off-target effects, IP challenges, regulatory uncertainty in some regions

Table 2.
Summary on traditional genetic engineering and CRISPR-mediated genome editing for food and nutritional enhancement [16, 84].

public-sector researchers. Addressing these concerns through transparent regulation, inclusive innovation, and open-access biotechnologies is essential for equitable and sustainable adoption [107, 108]. The summary of traditional genetic engineering technology and CRISPR-mediated genome editing for food and nutritional security is depicted in **Table 2**.

6. Future perspectives

The forthcoming crop genetic engineering is being rapidly altered by the combination of multi-omics technologies (genomics, transcriptomics, proteomics, metabolomics, and epigenomics) and artificial intelligence (AI), permitting more accurate and prognostic methods for crop improvement [109, 110]. These platforms simplify systems-level consideration of plant responses to environmental stimuli and recognize key molecular targets for nutritional quality, yield enhancement, and stress tolerance [111]. AI-powered tools like machine learning algorithms are now used to model genotypic and phenotypic relations, predict gene function, and accelerate genome editing approaches [112]. Gene stacking is the strategic incorporation of multiple genes conferring complementary traits has emerged as a promising approach for developing next-generation crops capable of thriving across diverse agroecological conditions [8]. Together with precision agriculture which utilizes sensors, robotics, and data analytics for real-time crop monitoring and optimized input management these innovations provide a sustainable pathway to maximize productivity while minimizing ecological impact [113]. However, understanding the full possibility of these tools necessitates worldwide collaboration in research, open data sharing, harmonized policy reforms, and capacity building in developing nations [114]. International initiatives such as the Consultative Group on International Agricultural Research (CGIAR) Research Programs, the FAO's International Treaty on Plant Genetic Resources, and CRISPR regulatory harmonization efforts are dangerous for reasonable access to novelties and addressing food and nutritional security under fluctuating climatic circumstances [115]. As bioengineering converges with digital agriculture, interdisciplinary collaboration and science-based policies will be key to shaping a resilient, inclusive, and sustainable agricultural future [116].

Modern genetic engineering for crop nutrition rapidly evolves with advanced technologies that offer unprecedented precision and adaptability [77]. Among these, precision epigenome editing has emerged as a powerful strategy for modulating the expression of key genes involved in nutrient biosynthesis without altering the DNA sequence [117]. CRISPR/dCas9-based tools, fused with histone acetyltransferases or demethylases, have been employed to upregulate genes involved in iron transport (e.g., *IRT1*, *NAS* genes) or vitamin E biosynthesis, enhancing nutrient availability in crops without traditional transgenic approaches [118]. Synthetic promoters further augment nutritional trait development by enabling precise spatiotemporal control of gene expression [119]. Custom promoters can be engineered to drive nutrient-specific gene expression in edible tissues, such as enhancing β-carotene biosynthesis in rice endosperm or iron accumulation in wheat grain while avoiding pleiotropic effects in vegetative tissues [120]. These promoters allow fine-tuned expression of biosynthetic genes under specific developmental or environmental cues, improving efficacy and biosafety. Although gene drives are not yet widely used for nutritional traits, theoretical frameworks suggest they could spread beneficial alleles such as those enhancing micronutrient content within wild or landrace populations, especially for traits controlled by recessive alleles or requiring high fixation [121]. For instance, native varieties could be enhanced to produce higher provitamin A or folate levels without continuous breeding interventions. Finally, AI-driven CRISPR design is transforming the precision of genome editing for nutritional traits. Machine learning tools like DeepCRISPR or CRISPR-GE now predict highly efficient and specific guide RNAs for editing biosynthetic pathway genes, minimizing off-target effects [122]. These tools are already being used to improve lysine and methionine content in cereals, tomato flavonoid profiles, and rice iron transport [123]. AI also helps prioritize targets by analyzing multi-omics data, improving the predictability and success rate of editing for nutritional enhancement [124]. When integrated with systems biology and precision breeding, these emerging tools hold enormous potential to improve the nutritional profile, bioavailability, and tissue-specific accumulation of key nutrients in staple crops. Their responsible implementation could significantly contribute to eradicating micronutrient deficiencies and advancing global food and nutritional security.

7. Conclusion

Genetic engineering technology addresses global food and nutritional insecurity by enhancing stress-tolerant capacity and nutrient content and expanding crops with high yields. By modifying plant genomes, researchers have developed biofortified crops such as Golden Rice and iron-rich beans that combat hidden hunger and micronutrient deficiencies in susceptible populations. Moreover, genetically modified plants contribute to climate resilience and sustainable agriculture by decreasing reliance on agrochemicals and improving resource use and yield. Despite these advances, environmental risks, biosafety regulations, ethical concerns, and public perception must be addressed through apparent ascendency, science- and technology-based strategies, and comprehensive discussion. Developing technologies like CRISPR/Cas systems, identification of traits via omics technology, and AI-guided crop design have the potential to speed up modernization. To fully realize the benefits of genetic engineering, global partnerships, equitable access to biotechnological tools, and farmer-centered approaches are essential for building a resilient and nutritionally secure future.

Conflict of interest

The authors declared that there is no conflict of interest in this manuscript.

Author details

Muthukumar Elankumaran[1], Anbuselvan Antony[2], Shivaanie Nandakumar[3], Vidyasagar Pari[4], Prosenjeet Chakraborthy[5], Sundarasamy Dhanapal[6], Chinnappan Deepa Joan of Arc[6] and Krishnagowdu Saravanan[6]*

1 Department of Botany, Bharathiar University, Coimbatore, Tamil Nadu, India

2 Department of Plant Pathology, Centre for Plant Protection Studies, Tamil Nadu Agricultural University, Coimbatore, Tamil Nadu, India

3 Department of Botany, PSGR Krishnammal College for Women, Coimbatore, Tamil Nadu, India

4 Department of Microbiology, Sri Venkateswara University, Tirupathi, Andra Pradesh, India

5 Department of Molecular Biosciences, Swami Vivekananda Yoga Anusandhana Samsthana University, Bangalore, Karnataka, India

6 Post Graduate and Research Centre in Biotechnology, Arignar Anna College (Arts and Science), Krishnagiri, Tamil Nadu, India

*Address all correspondence to: saravanankrishnagowdu95@gmail.com

References

[1] Tulchinsky TH. Micronutrient deficiency conditions: Global health issues. Public Health Reviews. 2010;**32**:243-255. DOI: 10.1007/BF03391600

[2] Graham RD, Welch RM, Saunders DA, Ortiz-Monasterio I, Bouis HE, Bonierbale M, et al. Nutritious Subsistence Food Systems. 2007:1-74. DOI: 10.1016/S0065-2113(04)92001-9

[3] Bouis HE, Saltzman A. Improving nutrition through biofortification: A review of evidence from HarvestPlus, 2003 through 2016. Global Food Security. 2017;**12**:49-58. DOI: 10.1016/j.gfs.2017.01.009

[4] Stuthman DD, Leonard KJ, Miller-Garvin J. Breeding crops for durable resistance to disease. Advances in Agronomy. 2007;**95**:319-367

[5] Benavente E, Giménez E. Modern approaches for the genetic improvement of rice, wheat and maize for abiotic constraints-related traits: A comparative overview. Agronomy. 2021;**11**(2):376

[6] Mastrangelo AM, Mazzucotelli E, Guerra D, De Vita P, Cattivelli L. Improvement of drought resistance in crops: From conventional breeding to genomic selection. In: Crop Stress and its Management: Perspectives and Strategies. Dordrecht: Springer Netherlands; 2011. pp. 225-259

[7] Dwivedi SL, Garcia-Oliveira AL, Govindaraj M, Ortiz R. Biofortification to avoid malnutrition in humans in a changing climate: Enhancing micronutrient bioavailability in seed, tuber, and storage roots. Frontiers in Plant Science. 2023;**14**. DOI: 10.3389/fpls.2023.1119148

[8] Khan AA, Iqbal B, Jalal A, Khan KA, Al-Andal A, Khan I, et al. Advanced molecular approaches for improving crop yield and quality: A review. Journal of Plant Growth Regulation. 2024;**43**:2091-2103. DOI: 10.1007/s00344-024-11253-7

[9] Chen K, Wang Y, Zhang R, Zhang H, Gao C. CRISPR/Cas genome editing and precision plant breeding in agriculture. Annual Review of Plant Biology. 2019;**70**:667-697. DOI: 10.1146/annurev-arplant-050718-100049

[10] Jha SK, McDermott J, Bacon G, Lannon C, Joshi PK, Dubé L. Convergent innovation for affordable nutrition, health, and health care: The global pulse roadmap. Annals of the New York Academy of Sciences. 2014;**1331**:142-156. DOI: 10.1111/nyas.12543

[11] Simkin AJ. Genetic engineering for global food security: Photosynthesis and biofortification. Plants. 2019;**8**:586. DOI: 10.3390/plants8120586

[12] González Guzmán M, Cellini F, Fotopoulos V, Balestrini R, Arbona V. New approaches to improve crop tolerance to biotic and abiotic stresses. Physiologia Plantarum. 2022;**174**. DOI: 10.1111/ppl.13547

[13] Gelvin SB. Integration of agrobacterium T-DNA into the plant genome. Annual Review of Genetics. 2017;**51**:195-217. DOI: 10.1146/annurev-genet-120215-035320

[14] Liu G, Campbell BC, Godwin ID. Sorghum genetic transformation by particle bombardment. In: Henry R, Furtado A, editors. Cereal Genomics. Methods in Molecular Biology. Vol. 1099. Totowa, NJ: Humana Press; 2014. DOI: 10.1007/978-1-62703-715-0_18

[15] Singh OV, Ghai S, Paul D, Jain RK. Genetically modified crops: Success, safety assessment, and public concern. Applied Microbiology and Biotechnology. 2006;**71**:598-607. DOI: 10.1007/s00253-006-0449-8

[16] Jaganathan D, Ramasamy K, Sellamuthu G, Jayabalan S, Venkataraman G. CRISPR for crop improvement: An update review. Frontiers in Plant Science. 2018;**9**. DOI: 10.3389/fpls.2018.00985

[17] Saravanan K, Praveenkumar K, Vidya N, Gowtham K, Saravanan M. Enhancement of agricultural crops: A CRISPR/Cas9-based approach. In: Vegetable Crops - Health Benefits and Cultivation. London, UK: IntechOpen; 2022. DOI: 10.5772/intechopen.100641

[18] Kandwal P, Fujiwara T, Kamiya T. OsVIT2 mutation increases Fe and Zn of grain without compromising the growth in paddy field. Frontiers in Plant Science. 2022;**13**. DOI: 10.3389/fpls.2022.868661

[19] Haun W, Coffman A, Clasen BM, Demorest ZL, Lowy A, Ray E, et al. Improved soybean oil quality by targeted mutagenesis of the fatty acid desaturase 2 gene family. Plant Biotechnology Journal. 2014;**12**:934-940. DOI: 10.1111/pbi.12201

[20] Sahu A, Verma R, Gupta U, Kashyap S, Sanyal I. An overview of targeted genome editing strategies for reducing the biosynthesis of phytic acid: An anti-nutrient in crop plants. Molecular Biotechnology. 2024;**66**:11-25. DOI: 10.1007/s12033-023-00722-1

[21] Shahzad R, Jamil S, Ahmad S, Nisar A, Khan S, Amina Z, et al. Biofortification of cereals and pulses using new breeding techniques: Current and future perspectives. Frontiers in Nutrition. 2021;**8**. DOI: 10.3389/fnut.2021.721728

[22] Miyagishi M, Taira K. Development and application of siRNA expression vector. Nucleic Acids Research. 2002;**Suppl**:113-114. DOI: 10.1093/nass/2.1.113

[23] Aggarwal S, Kumar A, Bhati KK, Kaur G, Shukla V, Tiwari S, et al. RNAi-mediated downregulation of inositol pentakisphosphate kinase (IPK1) in wheat grains decreases phytic acid levels and increases Fe and Zn accumulation. Frontiers in Plant Science. 2018;**9**. DOI: 10.3389/fpls.2018.00259

[24] Sunilkumar G, Campbell LM, Puckhaber L, Stipanovic RD, Rathore KS. Engineering cottonseed for use in human nutrition by tissue-specific reduction of toxic gossypol. Proceedings of the National Academy of Sciences. 2006;**103**:18054-18059. DOI: 10.1073/pnas.0605389103

[25] Cagliari D, Dias NP, Galdeano DM, dos Santos EÁ, Smagghe G, Zotti MJ. Management of pest insects and plant diseases by non-transformative RNAi. Frontiers in Plant Science. 2019;**10**. DOI: 10.3389/fpls.2019.01319

[26] Rollié S, Mangold M, Sundmacher K. Designing biological systems: Systems engineering meets synthetic biology. Chemical Engineering Science. 2012;**69**:1-29. DOI: 10.1016/j.ces.2011.10.068

[27] Mipeshwaree Devi A, Khedashwori Devi K, Premi Devi P, Lakshmipriyari Devi M, Das S. Metabolic engineering of plant secondary metabolites: Prospects and its technological challenges. Frontiers in Plant Science. 2023;**14**. DOI: 10.3389/fpls.2023.1171154

[28] Butelli E, Titta L, Giorgio M, Mock H-P, Matros A, Peterek S, et al. Enrichment of tomato fruit with health-promoting anthocyanins by expression

of select transcription factors. Nature Biotechnology. 2008;**26**:1301-1308. DOI: 10.1038/nbt.1506

[29] De Steur H, Blancquaert D, Strobbe S, Lambert W, Gellynck X, Van Der Straeten D. Status and market potential of transgenic biofortified crops. Nature Biotechnology. 2015;**33**:25-29. DOI: 10.1038/nbt.3110

[30] Ruiz-Lopez N, Haslam RP, Napier JA, Sayanova O. Successful high-level accumulation of fish oil omega-3 long-chain polyunsaturated fatty acids in a transgenic oilseed crop. The Plant Journal. 2014;**77**:198-208. DOI: 10.1111/tpj.12378

[31] Leal-Alves C, Deng Z, Kermeci N, Shih SCC. Integrating microfluidics and synthetic biology: Advancements and diverse applications across organisms. Lab on a Chip. 2024;**24**:2834-2860. DOI: 10.1039/D3LC01090B

[32] Ofori KF, Antoniello S, English MM, Aryee ANA. Improving nutrition through biofortification—A systematic review. Frontiers in Nutrition. 2022;**9**. DOI: 10.3389/fnut.2022.1043655

[33] Buturi CV, Mauro RP, Fogliano V, Leonardi C, Giuffrida F. Mineral biofortification of vegetables as a tool to improve human diet. Food. 2021;**10**:223. DOI: 10.3390/foods10020223

[34] Zhu C, Sanahuja G, Yuan D, Farré G, Arjó G, Berman J, et al. Biofortification of plants with altered antioxidant content and composition: Genetic engineering strategies. Plant Biotechnology Journal. 2013;**11**:129-141. DOI: 10.1111/j.1467-7652.2012.00740.x

[35] Nayak S, Mandi S, Baral K, Prasanna R, Shivay YS. Agronomic approaches for biofortification of staple food crops. In: Biofortification Staple Crop. Singapore: Springer Singapore; 2022. pp. 483-517. DOI: 10.1007/978-981-16-3280-8_19

[36] Miller BDD, Welch RM. Food system strategies for preventing micronutrient malnutrition. Food Policy. 2013;**42**:115-128. DOI: 10.1016/j.foodpol.2013.06.008

[37] Díaz-Gómez J, Twyman RM, Zhu C, Farré G, Serrano JC, Portero-Otin M, et al. Biofortification of crops with nutrients: Factors affecting utilization and storage. Current Opinion in Biotechnology. 2017;**44**:115-123. DOI: 10.1016/j.copbio.2016.12.002

[38] Phorbee O, Olatunde G, Aderonmu D, Ikerionwu D, Ojo A, Oguzor G, et al. Appropriate post-harvest technologies for biofortified crops pro enhanced utilization, value addition, and micronutrient retention. In: New Advances in Postharvest Technology. London, UK: IntechOpen; 2023. DOI: 10.5772/intechopen.110473

[39] Banerjee S, Roy P, Nandi S, Roy S. Advanced biotechnological strategies towards the development of crops with enhanced micronutrient content. Plant Growth Regulation. 2023;**100**:355-371. DOI: 10.1007/s10725-023-00968-4

[40] Prasanna R, Nain L, Rana A, Shivay YS. Biofortification with microorganisms: Present status and future challenges. In: Biofortification Food Crop. New Delhi: Springer India; 2016. pp. 249-262. DOI: 10.1007/978-81-322-2716-8_19

[41] Fanzo J. Strengthening the engagement of food and health systems to improve nutrition security: Synthesis and overview of approaches to address malnutrition. Global Food

Security. 2014;**3**:183-192. DOI: 10.1016/j.gfs.2014.09.001

[42] Rein MJ, Renouf M, Cruz-Hernandez C, Actis-Goretta L, Thakkar SK, da Silva Pinto M. Bioavailability of bioactive food compounds: A challenging journey to bioefficacy. British Journal of Clinical Pharmacology. 2013;**75**:588-602. DOI: 10.1111/j.1365-2125.2012.04425.x

[43] Paine JA, Shipton CA, Chaggar S, Howells RM, Kennedy MJ, Vernon G, et al. Improving the nutritional value of Golden Rice through increased pro-vitamin A content. Nature Biotechnology. 2005;**23**:482-487. DOI: 10.1038/nbt1082

[44] Goto F, Yoshihara T, Shigemoto N, Toki S, Takaiwa F. Iron fortification of rice seed by the soybean ferritin gene. Nature Biotechnology. 1999;**17**:282-286. DOI: 10.1038/7029

[45] Johnson AAT, Kyriacou B, Callahan DL, Carruthers L, Stangoulis J, Lombi E, et al. Constitutive overexpression of the OsNAS gene family reveals single-gene strategies for effective iron- and zinc-biofortification of rice endosperm. PLoS ONE. 2011;**6**:e24476. DOI: 10.1371/journal.pone.0024476

[46] Storozhenko S, De Brouwer V, Volckaert M, Navarrete O, Blancquaert D, Zhang G-F, et al. Folate fortification of rice by metabolic engineering. Nature Biotechnology. 2007;**25**:1277-1279. DOI: 10.1038/nbt1351

[47] Wakasa K. High-level tryptophan accumulation in seeds of transgenic rice and its limited effects on agronomic traits and seed metabolite profile. Journal of Experimental Botany. 2006;**57**:3069-3078. DOI: 10.1093/jxb/erl068

[48] Masuda H, Usuda K, Kobayashi T, Ishimaru Y, Kakei Y, Takahashi M, et al. Overexpression of the barley nicotianamine synthase gene HvNAS1 increases iron and zinc concentrations in rice grains. Rice. 2009;**2**:155-166. DOI: 10.1007/s12284-009-9031-1

[49] Ishimoto M, Rahman SM, Hanafy MS, Khalafalla MM, El-Shemy HA, Nakamoto Y, et al. Evaluation of amino acid content and nutritional quality of transgenic soybean seeds with high-level tryptophan accumulation. Molecular Breeding. 2010;**25**:313-326. DOI: 10.1007/s11032-009-9334-3

[50] Shintani D, DellaPenna D. Elevating the vitamin E content of plants through metabolic engineering. Science (80-.). 1998;**282**:2098-2100. DOI: 10.1126/science.282.5396.2098

[51] Arun M, Subramanyam K, Theboral J, Sivanandhan G, Rajesh M, Kapil Dev G, et al. Transfer and targeted overexpression of γ-tocopherol methyltransferase (γ-TMT) gene using seed-specific promoter improves tocopherol composition in Indian soybean cultivars. Applied Biochemistry and Biotechnology. 2014;**172**:1763-1776. DOI: 10.1007/s12010-013-0645-9

[52] Tavva VS, Kim Y-H, Kagan IA, Dinkins RD, Kim K-H, Collins GB. Increased α-tocopherol content in soybean seed overexpressing the *Perilla frutescens* γ-tocopherol methyltransferase gene. Plant Cell Reports. 2006;**26**:61-70. DOI: 10.1007/s00299-006-0218-2

[53] Yi J, Derynck MR, Li X, Telmer P, Marsolais F, Dhaubhadel S. A single-repeat MYB transcription factor, GmMYB176, regulates CHS8 gene expression and affects isoflavonoid biosynthesis in soybean. The Plant Journal. 2010;**62**:1019-1034. DOI: 10.1111/j.1365-313X.2010.04214.x

[54] Torabi S, Sukumaran A, Dhaubhadel S, Johnson SE, LaFayette P, Parrott WA, et al. Effects of type I diacylglycerol O-acyltransferase (DGAT1) genes on soybean (*Glycine max* L.) seed composition. Scientific Reports. 2021;**11**:2556. DOI: 10.1038/s41598-021-82131-5

[55] Eckert H, LaVallee B, Schweiger BJ, Kinney AJ, Cahoon EB, Clemente T. Co-expression of the borage Δ6 desaturase and the Arabidopsis Δ15 desaturase results in high accumulation of stearidonic acid in the seeds of transgenic soybean. Planta. 2006;**224**:1050-1057. DOI: 10.1007/s00425-006-0291-3

[56] Diretto G, Al-Babili S, Tavazza R, Papacchioli V, Beyer P, Giuliano G. Metabolic engineering of potato carotenoid content through tuber-specific overexpression of a bacterial mini-pathway. PLoS ONE. 2007;**2**:e350. DOI: 10.1371/journal.pone.0000350

[57] Welsch R, Arango J, Bär C, Salazar B, Al-Babili S, Beltrán J, et al. Provitamin A accumulation in cassava (Manihot esculenta) roots driven by a single nucleotide polymorphism in a phytoene synthase gene. The Plant Cell. 2010;**22**:3348-3356. DOI: 10.1105/tpc.110.077560

[58] Luo J, Butelli E, Hill L, Parr A, Niggeweg R, Bailey P, et al. AtMYB12 regulates caffeoyl quinic acid and flavonol synthesis in tomato: Expression in fruit results in very high levels of both types of polyphenol. The Plant Journal. 2008;**56**:316-326. DOI: 10.1111/j.1365-313X.2008.03597.x

[59] D'Ambrosio C, Giorio G, Marino I, Merendino A, Petrozza A, Salfi L, et al. Virtually complete conversion of lycopene into β-carotene in fruits of tomato plants transformed with the tomato lycopene β-cyclase (tlcy-b) cDNA. Plant Science. 2004;**166**:207-214. DOI: 10.1016/j.plantsci.2003.09.015

[60] Paul J, Khanna H, Kleidon J, Hoang P, Geijskes J, Daniells J, et al. Golden bananas in the field: Elevated fruit pro-vitamin A from the expression of a single banana transgene. Plant Biotechnology Journal. 2017;**15**:520-532. DOI: 10.1111/pbi.12650

[61] Lännenpää M. Heterologous expression of AtMYB12 in kale (*Brassica oleracea* var. *acephala*) leads to high flavonol accumulation. Plant Cell Reports. 2014;**33**:1377-1388. DOI: 10.1007/s00299-014-1623-6

[62] Adiaha MS, Agba OA, Attoe EE, Ojikpong TO, Kekong MA, Obio A, et al. Effect of maize (Zea mays L.) on human development and the future of man-maize survival: A review. World Scientific News. 2016;**59**:52-62

[63] Hossain F, Sarika K, Muthusamy V, Zunjare RU, Gupta HS. Quality protein maize for nutritional security. In: Quality Breeding in Field Crops. Cham: Springer International Publishing; 2019. pp. 217-237. DOI: 10.1007/978-3-030-04609-5_11

[64] Zhang S, Zhan J, Yadegari R. Maize opaque mutants are no longer so opaque. Plant Reproduction. 2018;**31**:319-326. DOI: 10.1007/s00497-018-0344-3

[65] Mishra SJ, Gopinath I, Muthusamy V, Zunjare RU, Chand G, Venkatesh AKT, et al. Unraveling the interactive effect of opaque2 and waxy1 genes on kernel nutritional qualities and physical properties in maize (*Zea mays* L.). Scientific Reports. 2025;**15**:3425. DOI: 10.1038/s41598-025-87666-5

[66] Huang S, Adams WR, Zhou Q, Malloy KP, Voyles DA, Anthony J, et al. Improving nutritional quality of

maize proteins by expressing sense and antisense zein genes. Journal of Agricultural and Food Chemistry. 2004;**52**:1958-1964. DOI: 10.1021/jf0342223

[67] Majumder S, Datta K, Datta SK. Rice biofortification: High iron, zinc, and vitamin-A to fight against "hidden hunger". Agronomy. 2019;**9**:803. DOI: 10.3390/agronomy9120803

[68] Rout GR, Sahoo S. Role of iron in plant growth and metabolism. Reviews in Agricultural Science. 2015;**3**:1-24. DOI: 10.7831/ras.3.1

[69] Roy C, Kumar S, Ranjan RD, Kumhar SR, Govindan V. Genomic approaches for improving grain zinc and iron content in wheat. Frontiers in Genetics. 2022;**13**. DOI: 10.3389/fgene.2022.1045955

[70] Nasim A, Hao J, Tawab F, Jin C, Zhu J, Luo S, et al. Micronutrient biofortification in wheat: QTLs, candidate genes and molecular mechanism. International Journal of Molecular Sciences. 2025;**26**:2178. DOI: 10.3390/ijms26052178

[71] Davies K, Espley R. Opportunities and challenges for metabolic engineering of secondary metabolite pathways for improved human health characters in fruit and vegetable crops. New Zealand Journal of Crop and Horticultural Science. 2013;**41**:154-177. DOI: 10.1080/01140671.2013.793730

[72] Sun W, Shahrajabian MH. Therapeutic potential of phenolic compounds in medicinal plants—Natural health products for human health. Molecules. 2023;**28**:1845. DOI: 10.3390/molecules28041845

[73] Bin-Jumah MN, Nadeem MS, Gilani SJ, Mubeen B, Ullah I, Alzarea SI, et al. Lycopene: A natural arsenal in the war against oxidative stress and cardiovascular diseases. Antioxidants. 2022;**11**:232. DOI: 10.3390/antiox11020232

[74] Syed RU, Moni SS, Bin Break MK, Khojali WMA, Jafar M, Alshammari MD, et al. Broccoli: A multi-faceted vegetable for health: An in-depth review of its nutritional attributes, antimicrobial abilities, and anti-inflammatory properties. Antibiotics. 2023;**12**:1157. DOI: 10.3390/antibiotics12071157

[75] Zhang Y-J, Gan R-Y, Li S, Zhou Y, Li A-N, Xu D-P, et al. Antioxidant phytochemicals for the prevention and treatment of chronic diseases. Molecules. 2015;**20**:21138-21156. DOI: 10.3390/molecules201219753

[76] Agostoni C, Baglioni M, La Vecchia A, Molari G, Berti C. Interlinkages between climate change and food systems: The impact on child malnutrition—Narrative review. Nutrients. 2023;**15**:416. DOI: 10.3390/nu15020416

[77] Khokhar Voytas A, Shahbaz M, Maqsood MF, Zulfiqar U, Naz N, Iqbal UZ, et al. Genetic modification strategies for enhancing plant resilience to abiotic stresses in the context of climate change. Functional & Integrative Genomics. 2023;**23**:283. DOI: 10.1007/s10142-023-01202-0

[78] Kumar A, Kerketta A, Dewali S, Sharma N, Kumari Panda A, Singh Bisht S. Tackling hidden hunger: Understanding micronutrient deficiency and effective mitigation strategies. In: Emerging Solutions in Sustainable Food and Nutrition Security. Cham: Springer International Publishing; 2023. pp. 305-319. DOI: 10.1007/978-3-031-40908-0_12

[79] Kiani AK, Dhuli K, Donato K, Aquilanti B, Velluti V, Matera G, et al.

Main nutritional deficiencies. Journal of Preventive Medicine and Hygiene. 2022;**63**:E93-E101. DOI: 10.15167/2421-4248/jpmh2022.63.2S3.2752

[80] Koç E, Karayiğit B. Assessment of biofortification approaches used to improve micronutrient-dense plants that are a sustainable solution to combat hidden hunger. Journal of Soil Science and Plant Nutrition. 2022;**22**:475-500. DOI: 10.1007/s42729-021-00663-1

[81] Dubock A. An overview of agriculture, nutrition and fortification, supplementation and biofortification: Golden Rice as an example for enhancing micronutrient intake. Agriculture and Food Security. 2017;**6**:59. DOI: 10.1186/s40066-017-0135-3

[82] Wairich A, Ricachenevsky FK, Lee S. A tale of two metals: Biofortification of rice grains with iron and zinc. Frontiers in Plant Science. 2022;**13**. DOI: 10.3389/fpls.2022.944624

[83] Nozoye T. The nicotianamine synthase gene is a useful candidate for improving the nutritional qualities and Fe-deficiency tolerance of various crops. Frontiers in Plant Science. 2018;**9**. DOI: 10.3389/fpls.2018.00340

[84] Sarkar M, Gupta J. Biotechnological interventions to mitigate the influence of climate alterations on ecosystem viability and human well-being. In: Unraveling New Frontiers and Advances in Bioinformatics. Singapore: Springer Nature Singapore; 2024. pp. 387-407. DOI: 10.1007/978-981-97-7123-3_17

[85] Shariatipour N, Heidari B. Genetic-based biofortification of staple food crops to meet zinc and iron deficiency-related challenges. In: Plant Micronutrients. Cham: Springer International Publishing; 2020. pp. 173-223. DOI: 10.1007/978-3-030-49856-6_8

[86] Ebert A. Potential of underutilized traditional vegetables and legume crops to contribute to food and nutritional security, income and more sustainable production systems. Sustainability. 2014;**6**:319-335. DOI: 10.3390/su6010319

[87] Majumdar R, Rajasekaran K, Cary JW. RNA interference (RNAi) as a potential tool for control of mycotoxin contamination in crop plants: Concepts and considerations. Frontiers in Plant Science. 2017;**8**. DOI: 10.3389/fpls.2017.00200

[88] Yu J, Peng P, Zhang X, Zhao Q, Zhy D, Sun X, et al. Seed-specific expression of a lysine rich protein sb401 gene significantly increases both lysine and total protein content in maize seeds. Molecular Breeding. 2004;**14**:1-7. DOI: 10.1023/B:MOLB.0000037990.23718.d6

[89] Mayer JE, Pfeiffer WH, Beyer P. Biofortified crops to alleviate micronutrient malnutrition. Current Opinion in Plant Biology. 2008;**11**:166-170. DOI: 10.1016/j.pbi.2008.01.007

[90] Lucca P, Hurrell R, Potrykus I. Fighting iron deficiency anemia with iron-rich rice. The Journal of the American College of Nutrition. 2002;**21**:184S-190S. DOI: 10.1080/07315724.2002.10719264

[91] Krasovskaia E, Just DR. Food, nutrition, and related policy issues: Evidence-based policy and the credibility crisis. Q Open. 2025;**5**(3):qoae013. DOI: 10.1093/qopen/qoae013

[92] Yadav V, Pal D, Poonia AK. A study on genetically engineered foods: Need, benefits, risk, and current knowledge. Cell Biochemistry and Biophysics.

2024;**82**:1931-1946. DOI: 10.1007/s12013-024-01390-x

[93] Winter G. Cultivation restrictions for genetically modified plants. European Journal of Risk Regulation. 2016;7:120-143. DOI: 10.1017/S1867299X00005444

[94] Mueller NG, Flachs A. Domestication, crop breeding, and genetic modification are fundamentally different processes: Implications for seed sovereignty and agrobiodiversity. Agriculture and Human Values. 2022;**39**:455-472. DOI: 10.1007/s10460-021-10265-3

[95] Spring ÚO. Genetically modified organisms: A threat for food security and risk for food sovereignty and survival. In: Brauch H et al., editors. Coping with Global Environmental Change, Disasters and Security. Hexagon Series on Human and Environmental Security and Peace. Vol. 5. Berlin, Heidelberg: Springer; 2011. DOI: 10.1007/978-3-642-17776-7_62

[96] Das D, Singha DL, Paswan RR, Chowdhury N, Sharma M, Reddy PS, et al. Recent advancements in CRISPR/Cas technology for accelerated crop improvement. Planta. 2022;**255**:109. DOI: 10.1007/s00425-022-03894-3

[97] Zhang L, He W, Fu R, Wang S, Chen Y, Xu H. Guide-specific loss of efficiency and off-target reduction with Cas9 variants. Nucleic Acids Research. 2023;**51**:9880-9893. DOI: 10.1093/nar/gkad702

[98] Manghwar H, Li B, Ding X, Hussain A, Lindsey K, Zhang X, et al. CRISPR/Cas systems in genome editing: Methodologies and tools for sgRNA design, off-target evaluation, and strategies to mitigate off-target effects. Advancement of Science. 2020;7. DOI: 10.1002/advs.201902312

[99] Wang J, Zhang X, Cheng L, Luo Y. An overview and metanalysis of machine and deep learning-based CRISPR gRNA design tools. RNA Biology. 2020;**17**:13-22. DOI: 10.1080/15476286.2019.1669406

[100] Matongera N, Ndhlela T, van Biljon A, Kamutando CN, Labuschagne M. Combining ability and testcross performance of multi-nutrient maize under stress and non-stress environments. Frontiers in Plant Science. 2023;**14**. DOI: 10.3389/fpls.2023.1070302

[101] Li W, Huang C, Chen J. The application of CRISPR/Cas mediated gene editing in synthetic biology: Challenges and optimizations. Frontiers in Bioengineering and Biotechnology. 2022;**10**. DOI: 10.3389/fbioe.2022.890155

[102] Yang C, Zhang Y, Huang C. Reduction in cadmium accumulation in japonica rice grains by CRISPR/Cas9-mediated editing of OsNRAMP5. Journal of Integrative Agriculture. 2019;**18**:688-697. DOI: 10.1016/S2095-3119(18)61904-5

[103] Hassoun A, Cropotova J, Trif M, Rusu AV, Bobiş O, Nayik GA, et al. Consumer acceptance of new food trends resulting from the fourth industrial revolution technologies: A narrative review of literature and future perspectives. Frontiers in Nutrition. 2022;**9**. DOI: 10.3389/fnut.2022.972154

[104] Jin S, Li W, Naab FZ, Coles D, Frewer LJ. Consumer attitudes toward novel agrifood technologies: A critical review on genetic modification and synthetic biology. In: Present Knowledge in Food Safety. Amsterdam, Netherlands: Elsevier; 2023. pp. 1004-1014. DOI: 10.1016/B978-0-12-819470-6.00035-4

[105] Turnbull C, Lillemo M, Hvoslef-Eide TAK. Global regulation of

genetically modified crops amid the gene edited crop boom—A review. Frontiers in Plant Science. 2021;**12**. DOI: 10.3389/fpls.2021.630396

[106] Schwartz MK, Dunn SL, Gendron WAC, Helm JE, Kamau WS, Mark-Shadbolt M, et al. Principles for introducing new genes and species for conservation. Trends in Ecology & Evolution. 2025;**40**:296-307. DOI: 10.1016/j.tree.2024.11.011

[107] Ibraimov K, Umaralieva M, AbduganievaS, ShamshetovaA, SharipovA, Safarova E. Bio-tech innovations: Bridging biology and technology for inclusive support. BIO Web of Conferences. 2024;**141**:04004. DOI: 10.1051/bioconf/202414104004

[108] Singh K, Zeeshan Z, Karna S. The regulation of biotechnology: Ethical, legal, and social implications. Interdisciplinary Studies in Society, Law, and Politics. 2023;**2**:44-50. DOI: 10.61838/kman.isslp.2.3.6

[109] Fan B-L, Chen L-H, Chen L-L, Guo H. Integrative multi-omics approaches for identifying and characterizing biological elements in crop traits: Current progress and future prospects. International Journal of Molecular Sciences. 2025;**26**:1466. DOI: 10.3390/ijms26041466

[110] Zhang Y, Huang G, Zhao Y, Lu X, Wang Y, Wang C, et al. Revolutionizing crop breeding: Next-generation artificial intelligence and big data-driven intelligent design. Engineering. 2025;**44**:245-255. DOI: 10.1016/j.eng.2024.11.034

[111] Saleem MH, Noreen S, Ishaq I, Saleem A, Khan KA, Ercisli S, et al. Omics technologies: Unraveling abiotic stress tolerance mechanisms for sustainable crop improvement. Journal of Plant Growth Regulation. 2025. DOI: 10.1007/s00344-025-11674-y

[112] Wójcik-Gront E, Zieniuk B, Pawełkowicz M. Harnessing AI-powered genomic research for sustainable crop improvement. Agriculture. 2024;**14**:2299. DOI: 10.3390/agriculture14122299

[113] Getahun S, Kefale H, Gelaye Y. Application of precision agriculture technologies for sustainable crop production and environmental sustainability: A systematic review. Scientific World Journal. 2024;**2024**. DOI: 10.1155/2024/2126734

[114] Linders D. Towards open development: Leveraging open data to improve the planning and coordination of international aid. Government Information Quarterly. 2013;**30**:426-434. DOI: 10.1016/j.giq.2013.04.001

[115] Akinbo O, Nkhabindze B, Amedu J, Ebegba R, Asagbra A, Ratemo BO, et al. Africa and zero hunger agenda: Genome editing policy landscape, challenges and opportunities. Frontiers in Bioengineering and Biotechnology. 2025;**13**. DOI: 10.3389/fbioe.2025.1526851

[116] Khan N. Unlocking innovation in crop resilience and productivity: Breakthroughs in biotechnology and sustainable farming. Innovation Discovery. 2024;**1**

[117] Laufer BI, Singh SM. Strategies for precision modulation of gene expression by epigenome editing: An overview. Epigenetics & Chromatin. 2015;**8**:34. DOI: 10.1186/s13072-015-0023-7

[118] Karlson CKS, Mohd-Noor SN, Nolte N, Tan BC. CRISPR/dCas9-based systems: Mechanisms and applications in plant sciences. Plants. 2021;**10**:2055. DOI: 10.3390/plants10102055

[119] Yasmeen E, Wang J, Riaz M, Zhang L, Zuo K. Designing artificial synthetic promoters for accurate, smart, and versatile gene expression in plants. Plant Communications. 2023;**4**:100558. DOI: 10.1016/j.xplc.2023.100558

[120] Bhat IA, Guleria K, Fayaz M, Roof-Ul-Qadir TA, Wani J, Qadir ZAK. Role of metabolic engineering in enhancing crop nutritional quality. In: Metabolic Engineering in Plants. Singapore: Springer Nature Singapore; 2022. pp. 145-170. DOI: 10.1007/978-981-16-7262-0_6

[121] Pandey B, Prakash P, Chandra Verma P, Srivastava R. Regulated gene expression by synthetic modulation of the promoter architecture in plants. In: Current Developments in Biotechnology and Bioengineering. Amsterdam, Netherlands: Elsevier; 2019. pp. 235-255. DOI: 10.1016/B978-0-444-64085-7.00010-1

[122] Kaul T, Verma R, Sony SK, Bharti J, Motelb KFA, Thangaraj AP, et al. CRISPR/Cas enables the remodeling of crops for sustainable climate-smart agriculture and nutritional security. In: Global Climate Change and Plant Stress Management. Hoboken, NJ, USA: Wiley; 2023. pp. 71-111. DOI: 10.1002/9781119858553.ch9

[123] Hall RD, D'Auria JC, Silva Ferreira AC, Gibon Y, Kruszka D, Mishra P, et al. High-throughput plant phenotyping: A role for metabolomics? Trends in Plant Science. 2022;**27**:549-563. DOI: 10.1016/j.tplants.2022.02.001

[124] Agarwal N, Nupur PK, Paul SKM. Artificial intelligence and machine learning for analysis of multi-omics. In: Multi-Omics Analysis of the Human Microbiome. Singapore: Springer Nature Singapore; 2024. pp. 339-354. DOI: 10.1007/978-981-97-1844-3_16

Chapter 6

Perspective Chapter: Seed Priming – A Novel Approach to Combat Microgravity Associated Stress Management in Plants

Kiran R. Kharat and Raveendran Pottathil

Abstract

The colonization of space presents a significant challenge due to the difficulty of cultivating crops in microgravity environments. For long-term space exploration and extraterrestrial settlements to become viable, establishing reliable methods of food production beyond Earth is critical. However, the effects of spatial factors on plant physiology—and their subsequent impact on crop yield—remain incompletely understood. Gravity is a fundamental regulator of plant development, influencing essential cellular processes such as cell division and growth. Alterations in gravitational conditions can profoundly affect both developmental pathways and physiological behaviors in plants. Notably, plant hormones exhibit differential responses to microgravity. For instance, cytokinin displays a distinct distribution pattern in microgravity, contrasting with auxin, which orchestrates directional root growth through a gravity-dependent signaling network. Emerging research has identified significant changes in gene expression under microgravity conditions, particularly in genes associated with the tricarboxylic acid (TCA) cycle, oxidative stress, carbohydrate metabolism, plant defense mechanisms, and DNA methylation (especially at the 5-position of cytosine). Short-term spaceflight experiments in low-Earth orbit have demonstrated disruptions in plant redox balance, stress-related protein expression, energy and amino acid metabolism, hormonal activity, and transcriptional regulation. Intriguingly, seeds primed prior to microgravity exposure exhibit enhanced resilience to space-induced stress, suggesting a promising strategy to improve plant adaptability in extraterrestrial environments. These findings underscore the need for further research to optimize agricultural systems for space colonization.

Keywords: plant growth in space, microgravity, seeds priming, stress response, gene expression

1. Introduction

Fresh sources of water, minerals, and human habitation can be found in space [1–3]. To sustain space travel and colonization, the ability to generate food in extraterrestrial settings is critical. Plants evolved on Earth with the help of gravity, photoperiod, barometric pressure, and other signals, and plants that live on space stations or other planets would face stressors that have never been seen before [1–4]. Plants are essential components of life support systems that are required to keep humans alive during prolonged space travel and on exoplanets. Plants thrive in space stations and have potential as food and energy sources. These plants may yield low quality or quantity of food under microgravity stress conditions [5–7]. The primary model organisms for investigating space life science over the last few decades have been rice, Arabidopsis, corn, fruit flies, and mice [8, 9]. Vegetables could be used by humans as a dependable food source on long-duration space missions. Previous research has shown that plants may modify their metabolic networks to adapt to space flight. The phytochemical components of fresh vegetables that contribute to the reduction of risks are essential for such applications [9, 10]. You need to be knowledgeable about how plants grow and develop in low gravity for space farming to be successful. Because gravity is one of the key factors that distinguish the environment on Earth from that in space or on exoplanets, it is also crucial for plant space biology [10].

Plants have not evolved mechanisms to cope with changes in gravity. To gain insight into how they respond to spaceflight conditions, researchers have investigated the biological pathways and processes that plants engage in while in orbit. Most studies exploring the effects of microgravity on plant growth have concentrated on specific responses, like gravitropism and gravimorphogenesis. In contrast, relatively few investigations have attempted to confirm whether plants can successfully complete a full seed-to-seed life cycle in space. Additionally, studies that have explored seed production in microgravity often pursued different objectives, such as evaluating seed quality or identifying the factors responsible for reproductive issues [11–13]. Due to the rising incidence of climate variability and extreme weather events, plants are more likely to experience multiple abiotic stresses throughout their lives rather than a single stress occurrence [14, 15]. To survive and adapt under such harsh conditions, plants need to develop effective mechanisms to cope with repeated stress exposures. Studies have shown that subjecting plants to mild biotic or abiotic stress beforehand can help them better endure more intense stress later, as demonstrated by Ramírez et al. [16] and Walter et al. [17]. This phenomenon, referred to as "plant priming," is linked to "plant stress memory" and has been found to enhance stress tolerance [15, 18]. While the term "plant priming" is commonly applied in cases involving biotic stress and chemical inducers, the same adaptive response is often called "hardening" or "acclimation" when related to abiotic stress [19–23]. This review explores recent insights into how priming influences plant development, both under Earth conditions and in microgravity.

1.1 Microgravity and plants

Studies on plants in space have a long history [14, 15, 18, 19]. Plants are regarded as essential tools for identifying gravity-related biological processes because of the link between plant architecture and gravity [19]. The completion of the plant cycle has additional space restrictions because of factors such as microgravity. The course of the cell cycle, mitotic index, and phase length were significantly altered in groundbreaking space and terrestrial research on lentil seedlings. There was variation in cell proliferation and growth rates between the Earth-grown and space-grown Arabidopsis seedlings [12, 13, 20–23]. In addition to plant-level priming, stress memory, and cross-stress resistance, seed priming plays a vital role in helping plants cope with adverse conditions. The process of imbibition marks the beginning of cell division and expansion and includes three phases of seed development: rapid water absorption, stabilization of water content, and continued water uptake [24].

1.2 Effect of microgravity on seed germination

The combined impact of multiple stressors is likely responsible for the diminished efficiency of various biological processes and the lower quality of plant structures formed in space. Among the primary challenges to plant development in space environments are the use of imbalanced nutrient solutions and lighting conditions that are suboptimal in both quality and intensity. These unfavorable environmental factors can impair plant growth directly and also heighten plant sensitivity to microgravity [25]. Research focusing on nutrient uptake in plants has revealed that absorption rates may increase, decrease, or remain unaffected, depending on factors such as the specific nutrient involved and the plant species [25–27].

In a recent review, Wolff et al. [27] concluded that although there are indications that plant nutrition is affected by the space environment, available information is limited and inconclusive [28, 29]. Merkys and Laurinavicius [30] used *Arabidopsis thaliana* in their first successful seed-to-seed experiment conducted in microgravity [22, 31]. Some viable seeds were observed, but the majority of seeds had nonviable embryos. *Brassica rapa* was the subject of a second successful experiment in space [32, 33]. Brassica seeds were found to have abnormal starch grain deposition, reduced protein, and fewer cotyledon cells, while being healthy and viable.

The challenging conditions of microgravity are more likely to account for the outcomes observed in both studies. Several researchers have reported changes in starch accumulation in a variety of plants grown in space- Arabidopsis, Lepidium roots, pepper plants, and maize root columellae [34–38]. A recurring issue in space-based plant research is the elevated production of ethylene, which can negatively impact plant growth. Experiments using clinostats [39] and those conducted in space have shown increased ethylene synthesis in plants [40]. Notably, Kuang et al. [41] reported exceptionally high ethylene concentrations (800–1200 ppb) aboard the Mir space station during a Brassica experiment. Remarkably, Brassica plants managed to produce seeds under these conditions, whereas wheat plants grown in the same environment failed to do so [42]. However, when seeds are bioprimed with plant growth-promoting *rhizobacteria* (PGPRs), the ethylene precursor 1-aminocyclopropane-1-carboxylic acid (ACC) is efficiently broken down by ACC deaminase, helping to regulate ethylene synthesis [43]. The researchers emphasize that achieving reproductive success in space requires not only a robust mother plant capable of supporting an energy-intensive process but also the successful completion of every stage of reproduction [25, 26].

The harsh conditions associated with microgravity—rather than the mere absence of gravity—are increasingly recognized as key contributors to the reduced efficiency of biological processes and the impaired quality of plant structures grown in space. Factors such as imbalanced nutrient media, suboptimal lighting, and heightened ethylene production present major stressors that compromise plant development and reproductive success in extraterrestrial environments [25, 34–38]. These stressors, individually and in combination, can drastically increase plant sensitivity to microgravity [25], challenging their ability to survive and complete their life cycle.

Ensuring strong initial growth—from dormant seed to viable seedling—is thus a critical step for completing the plant reproductive cycle in such resource-limited and stress-intensive environments [25, 44, 45]. While dry seeds are naturally resilient to adverse conditions, their resistance wanes quickly during germination, making the early seedling stage highly vulnerable to the types of physical and chemical stresses commonly encountered in space [25, 26]. Seedling survival under these challenging conditions hinges on three key attributes: rapid establishment in the substrate, the capacity to develop morpho-functional adaptations to cope with stress (including drought and biotic interactions), and a fast progression through early developmental stages. These traits, which are genetically determined, are heavily influenced by environmental stressors, such as those present in microgravity settings [26].

To counteract these effects, seed priming emerges as a vital strategy to support normal germination and seedling establishment under stress conditions, including those encountered in space. Research has demonstrated that hydropriming significantly improves germination rates and early seedling growth in maize (*Zea mays L.*) during drought stress [46], while hydroprimed cotton seeds (*Gossypium hirsutum L.*) exhibit enhanced growth and yield under limited water availability [47]. Under similar stress conditions, melatonin-primed rapeseeds show improved germination and early development. These findings highlight the potential of seed priming not only for terrestrial agriculture but also as a key tool to enhance plant performance in the extreme conditions of space.

1.3 Microgravity stress and breeding approaches

The microgravity environment in space disrupts several physiological and developmental processes in plants, presenting a unique stress condition not encountered on Earth. These disturbances include altered cell division, disrupted hormonal signaling, reduced nutrient uptake, and compromised reproductive development. Such stress not only affects plant morphology but also impairs yield and seed viability, crucial parameters for sustainable space farming [25, 38, 48].

To counteract these negative effects, plant breeding and biotechnological approaches are emerging as powerful tools. Traditional breeding, molecular breeding, and modern genome-editing techniques like CRISPR/Cas9 can be harnessed to develop stress-resilient genotypes capable of thriving under microgravity and other extraterrestrial stressors. These approaches focus on identifying and integrating traits that enhance germination, growth, and reproduction under space-relevant stress conditions. For example, identifying quantitative trait loci (QTLs) linked with traits such as efficient nutrient use, ethylene regulation, and enhanced antioxidant production could support the development of elite cultivars adapted for space. Similarly, marker-assisted selection (MAS) enables the rapid introgression of desirable traits, including enhanced stress memory and efficient hormonal responses, into target cultivars. Emerging omics technologies, including genomics, transcriptomics, proteomics, and metabolomics, have revealed stress-responsive gene networks and metabolic pathways triggered during spaceflight. These insights can guide the selection and engineering of crops with enhanced gravitropic insensitivity, stress hormone balance, and improved reproductive resilience [12, 13, 33, 43]. Furthermore, leveraging plant-microbe interactions through inoculation with beneficial microorganisms such as plant growth-promoting rhizobacteria (PGPRs) offers a promising avenue for improving plant performance in space. These microbes can mitigate the accumulation of ethylene by metabolizing its precursor ACC, enhancing root architecture and nutrient uptake under stress [43]. Combined with priming strategies, this represents a holistic approach to increasing plant robustness in extreme conditions.

Therefore, integrating breeding approaches with physiological priming and microbial inoculation can enable the design of multi-stress-tolerant plants, capable of completing their full life cycle and producing viable seeds in microgravity. This integrative strategy is essential for the success of bioregenerative life support systems on long-term space missions and planetary colonization efforts.

1.4 Jasmonic acid signaling and microgravity stress

Researchers at the University of Zurich have discovered that a specific plant hormone may be crucial for enabling plant growth in environments with limited nutrients and gravity, potentially paving the way for future extraterrestrial agriculture. The hormone strigolactone plays a key role in promoting symbiosis between plant roots and beneficial fungi, a relationship that enhances plant development and may prove effective even under extreme conditions, such as those found on the Moon. Secreted by plant roots, strigolactone supports mycorrhization, thereby boosting growth even in nutrient-poor soils. Unlike auxin, cytokinin, another plant hormone, exhibits distinct distribution patterns in microgravity compared to Earth. While auxin is known for its role in gravity sensing within root zones, cytokinins often act as regulatory counterparts to auxin [49, 50].

Hormonal priming has shown promise in improving plant performance under stress. For instance, treatments with auxin significantly enhanced wheat growth and led to greater grain yield compared to untreated plants [49–53]. Additional studies revealed that tomato seedlings treated with copper- and zinc-based elicitors experienced gene expression changes associated with the jasmonic acid (JA) signaling pathway, which is linked to stress responses such as pathogen attacks, hormone signaling, and mechanical injury. Notably, the Solyc12g009220.2 gene was upregulated in these treated plants [54]. Moreover, transcript levels of various auxin transporter genes, including Solyc01g007010.3, a RING-type E3 ubiquitin ligase, were significantly altered. Genes involved in the abscisic acid (ABA) signaling pathway were also found to be upregulated in primed seedlings, indicating the critical role of hormonal regulation in stress adaptation and development [54].

Genes associated with jasmonic acid-mediated signaling pathways play a crucial role in regulating plant defense mechanisms and responses to various stress factors (**Figure 1**). Specifically, genes involved in wound response and the regulation of stimulus response are modulated by jasmonic acid (**Figure 1**). These findings have potential implications for developing drought-resistant crop varieties. Enhancing plant resilience through stress-preconditioning treatments promotes healthy growth and may serve as a foundation for future research on plant priming and stress memory, particularly for enabling seed production under microgravity conditions.

1.5 Effect of microgravity on the citric acid cycle

Spaceflight alters the mitochondrial activity [30, 48] in plants. TCA flux is dramatically altered by numerous abiotic stressors [55]. In *Arabidopsis*, changes in TCA cycle-related genes have been validated using transcriptome analysis [24]. Because the adaptation of plants to abiotic stress is regulated by TCA, it is regarded as a crucial physiological response. Amino acids, fatty acids, and secondary metabolites can be synthesized from TCA cycle intermediates [25, 55]. According to our research, some of the acid content of organic acids include aconitic acid, alpha-ketoglutarate, aconitic acid, maleate, succinate, fumaric, fumuvic, pyruvic, citrate, and isocitric acid. Hell and Stephan [56] noted that numerous physiological and biochemical events rely on the presence of citric acid, which is an essential organic acid [26]. Citric acid has been

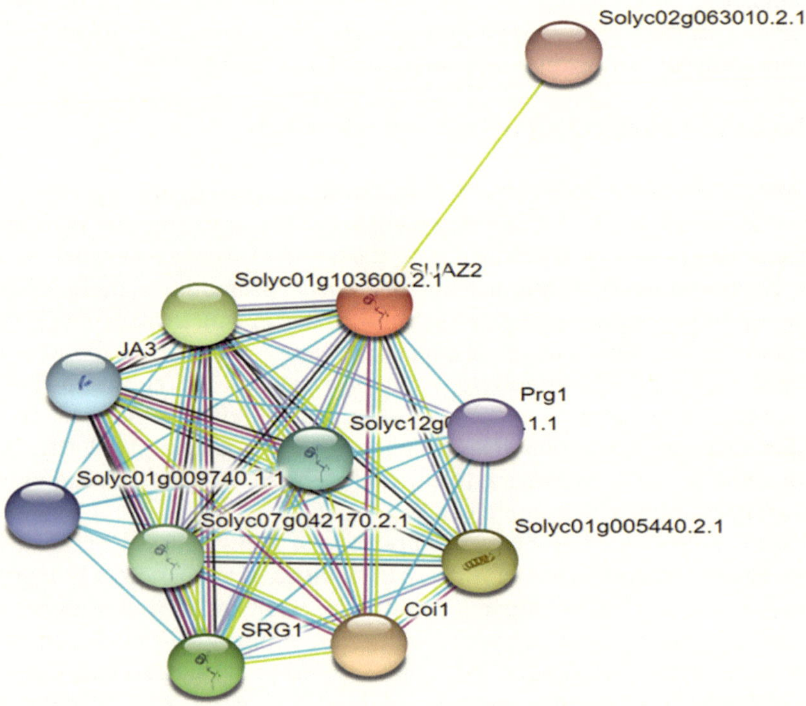

Biological Process (Gene Ontology)

GO-term	description	count in network	strength	false discovery rate
GO:2000022	Regulation of jasmonic acid mediated signaling pathway	7 of 25	2.95	5.32e-16
GO:0009611	Response to wounding	7 of 135	2.21	1.55e-11
GO:0031347	Regulation of defense response	7 of 194	2.06	1.23e-10
GO:0048583	Regulation of response to stimulus	8 of 768	1.52	9.07e-09
GO:0006950	Response to stress	9 of 3176	0.95	1.36e-05
GO:0050896	Response to stimulus	10 of 5588	0.75	6.43e-05
GO:0050794	Regulation of cellular process	10 of 5664	0.75	6.68e-05
				(less ...)

Figure 1.
String map of the JA3 (Jasmonic acid) with other pathways in Tomato. Where JA3 is Jasmonic acid linked to seven pathways for the regulation of various genes in response to stress or defense.

linked to aluminum poisoning, iron stress, heavy metal stress tolerance, and salt stress [27]. Citric acid level is a good indicator of the ability of rice cultivars to withstand spaceflight at different ages [48]. A comparable response has been observed in primed Arabidopsis seeds, where elevated levels of carboxylic acids have been shown to strengthen defense mechanisms through intricate signaling networks. Elevated levels of tricarboxylic acid (TCA) are related to an increase in plant resistance stimulated by inducers [57]. Similarly, Copper- and zinc-based elicitors have been shown to induce carboxylic acids in okra seeds [58].

1.6 Genes related to oxidative stress-related genes

Antioxidants and ROS-scavenging/detoxifying enzymes must maintain flexible redox states to mitigate oxidative damage [59–64]. During spaceflight, plants tend to express more ROS- and redox-associated genes compared to those grown under Earth's gravity. This upregulation reflects an adaptive response to the novel and potentially stressful environment of microgravity [32, 64–66]. Reactive oxygen species (ROS), produced in various subcellular compartments, elicit a wide range of physiological, biochemical, and molecular responses in plants [51, 61–64, 67–70]. To cope with the oxidative stress induced by microgravity, plants may engage multiple ROS-response mechanisms. This environmental unfamiliarity prompts plants, being sessile organisms, to deploy sensitive and adaptive strategies for metabolic regulation. Soluble sugars play a critical role in plant metabolism and structural integrity. In microgravity, sugar signaling interacts closely with ROS signaling to modulate the plant's redox state [71, 72]. While some aspects of sugar metabolism may adapt to spaceflight conditions—evident in crops like rice—the precise mechanisms can differ from terrestrial systems.

As nutrient and metabolite signaling molecules, soluble sugars can influence hormone signaling pathways, thereby affecting gene expression and protein abundance. Our previous study identified several significantly expressed genes involved in ROS scavenging under metal treatment, including prephenate/arogenate dehydrogenase (Solyc09g011870.2), Fe2OG dioxygenase (Solyc12g006370.2), and L-ascorbate oxidase (Solyc04g054690.3). These findings highlight the modulation of oxidative stress responses in tomato seedlings subjected to Cu and Zn treatments, which could offer insights into similar mechanisms under spaceflight-induced stress. In seeds, ROS act not only as damaging agents but also as signaling molecules during germination [73]. Oxidative modifications of storage proteins serve as triggers for reserve mobilization. For instance, eggplant seeds exhibited an increase in ROS levels after 24 hours of hydropriming, followed by a gradual decline [74]. Interestingly, ROS levels post-dry-back were even lower than in dry seeds, suggesting an effective antioxidant response and potential memory of oxidative stress, which could be relevant for seedling establishment in microgravity conditions [75].

1.7 Differential gene expression in plants in microgravity

Experiments conducted aboard the International Space Station (ISS) have demonstrated that spaceflight significantly alters gene expression patterns in plants. Although the full extent of these transcriptomic changes is still being unraveled, research indicates that DNA methylation at the fifth cytosine position, a critical epigenetic modification, plays an essential role in helping plants adapt to both terrestrial and extraterrestrial stress conditions. These findings suggest that exposure to spaceflight activates complex regulatory systems that could be leveraged to enhance plant resilience and performance beyond Earth. As more transcriptomic and proteomic data emerge from ISS-based plant studies, consistent trends are becoming evident. Genes that are differentially expressed during spaceflight influence a range of biological pathways, including redox regulation, stress response, energy and hormone metabolism, amino acid biosynthesis, and transcriptional control [9, 50, 65, 76–81]. While the ISS can support relatively normal plant development in orbit, this is only possible due to extensive metabolic and structural reprogramming induced by microgravity.

For instance, in *Arabidopsis thaliana*, microgravity-induced oxidative stress has been found to activate molecular chaperones, which protect cellular integrity [82]. Seeds of *Eruca sativa* that were exposed to 6 months of microgravity exhibited delayed growth and heightened sensitivity to senescence, along with reduced expression of genes linked to transcription and translation [83]. Similarly, seedlings of pea, maize, oat, mung bean, and sunflower grown in space showed impaired cell division, altered mitotic activity, and disrupted auxin transport in root regions. These observations confirm that spaceflight acts as a severe abiotic stressor, leading to broad genomic reprogramming that influences plant growth, development, and yield, as well as cellular structures like cell wall composition, as seen in *Arabidopsis*

[26, 84–86]. Parallel responses have also been documented in metal ion-primed tomato seedlings, where oxidative bursts triggered by copper and zinc-based elicitors resulted in the upregulation of WRKY transcription factors, including WRKY81. This particular transcription factor modulates jasmonic acid signaling and regulates genes involved in stress adaptation and biosynthesis (**Figure 2**).

These insights offer valuable direction for space agriculture breeding strategies. By identifying gene variants and regulatory sequences associated with traits such as improved oxidative stress resistance, optimized hormone signaling, and stable cell division, scientists can develop cultivars better suited for growth in microgravity. Moreover, integrating epigenetic and transcriptomic data into breeding pipelines—using tools like CRISPR-Cas9 and marker-assisted selection—could accelerate the creation of space-adapted crop varieties. Ultimately, the molecular insights gained from spaceflight experiments not only deepen our understanding of plant stress physiology but also provide a foundation for next-generation breeding aimed at supporting life beyond Earth.

Early gene expression studies indicated that the metabolic adaptation of plants to spaceflight relied on changes in calcium signaling, but they could not identify a single

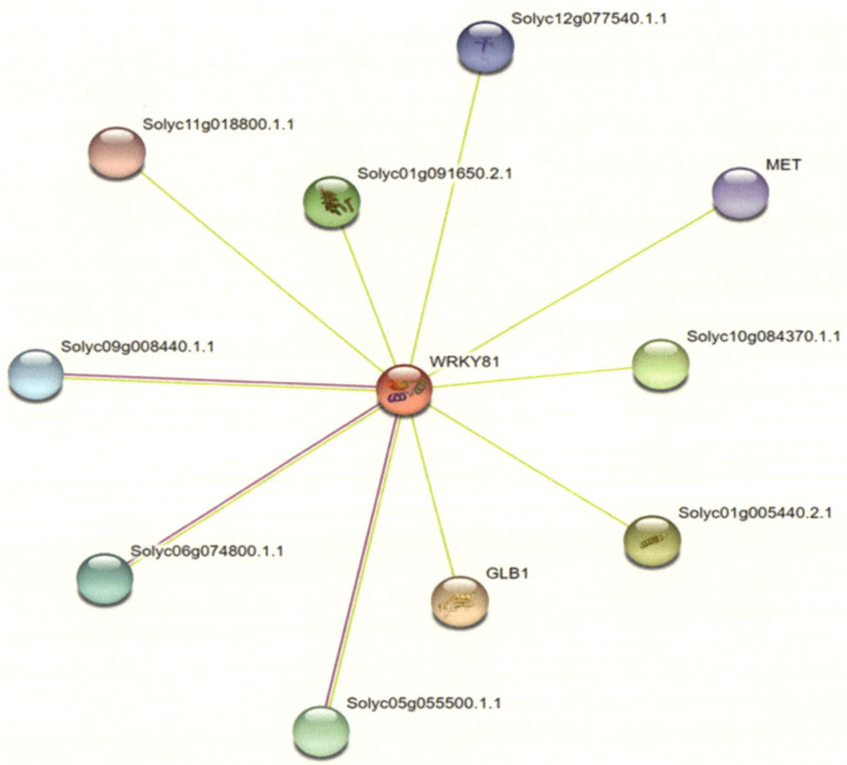

Biological Process (Gene Ontology)

GO-term	description	count in network	strength
GO:2000022	Regulation of jasmonic acid mediated signaling pathway	2 of 25	2.4
GO:0031326	Regulation of cellular biosynthetic process	7 of 2923	0.88
GO:0080090	Regulation of primary metabolic process	7 of 3525	0.8
GO:0050794	Regulation of cellular process	9 of 5664	0.7

Figure 2.
The WRKY related with other pathways in Tomato. Where, WRKY81 is transcription factor regulating the Jasmonic acid-mediated signaling pathway.

or specific environmental stress that characterized spaceflight [37–40, 42, 43]. In addition to the well-documented changes in gene expression, plants aboard the ISS also exhibit notable shifts in DNA methylation patterns. This supports the idea that plant adaptation to spaceflight involves epigenetic regulation, aligning with findings that gene regulatory networks are influenced by environmental stressors and ecological factors on Earth [87–96]. Numerous studies have demonstrated the central role of epigenetic modifications in regulating gene activity [97, 98] with DNA methylation emerging as a key mechanism by which organisms adjust to environmental stimuli through stable, heritable changes in gene expression. These modifications can become embedded in the genome and may be transmitted across generations [99–101]. Despite mounting evidence that DNA methylation is responsive to terrestrial environmental changes, few studies have explored this process in space. However, some research has examined plants grown on Earth from seeds previously exposed to spaceflight, as well as seedlings that experienced 60 hours of spaceflight aboard a satellite. These studies contribute to the growing body of work on spaceflight-associated epigenomic responses in plants [102]. Thanks to advancements in space-based research, well-replicated transcriptomic studies can now be performed aboard the ISS, offering deeper insights into the global transcriptional landscape of *Arabidopsis thaliana*. Transcriptomic data now span multiple cultivars, cell types, and growth conditions [103], revealing a substantial number of differentially expressed genes in space-grown plants. Interestingly, many of these genes encode proteins whose functions are still not fully understood, highlighting the need for further investigation [104].

2. Roots development in spaceflights and auxins

Beyond the well-established alterations in gene expression, plants grown aboard the ISS also demonstrate significant changes in DNA methylation patterns. This observation reinforces the idea that epigenetic regulation plays a key role in how plants adapt to spaceflight, consistent with the understanding that gene regulatory systems are influenced by environmental stressors and ecological conditions on Earth [105–107]. Extensive research has confirmed the importance of epigenetic mechanisms, particularly DNA methylation, in modulating gene expression [100–103]. DNA methylation enables organisms to respond to environmental cues by introducing stable and potentially heritable genetic modifications, which may be retained across generations [102–104]. Although there is substantial evidence supporting DNA methylation as a response to environmental factors on Earth, only a limited number of studies have examined this process in space conditions. Some of these studies include analyses of plants cultivated on Earth from spaceflight-exposed seeds, as well as seedlings subjected to 60 hours of microgravity during satellite missions. These findings are gradually expanding the field of space-related epigenomic research in plants [105].

With recent advancements in space-based research infrastructure, comprehensive and reproducible transcriptome analyses are now feasible aboard the ISS. These studies provide valuable insights into the transcriptional dynamics of *Arabidopsis thaliana* under spaceflight conditions. A growing body of transcriptomic data covers various plant cultivars, cell types, and environmental conditions [106], revealing numerous differentially expressed genes in space-grown plants. Notably, a significant portion of these genes encode proteins with yet-to-be-determined functions, indicating a promising area for future exploration [107].

IAA5 and IAA1 have been linked to several key developmental processes, including pollen maturation, lateral root initiation, auxin-mediated signaling, root branching, and gravitropic responses (**Figure 3**). Plant organ systems, shoot system development, and postembryonic development are also associated with IAA. The most

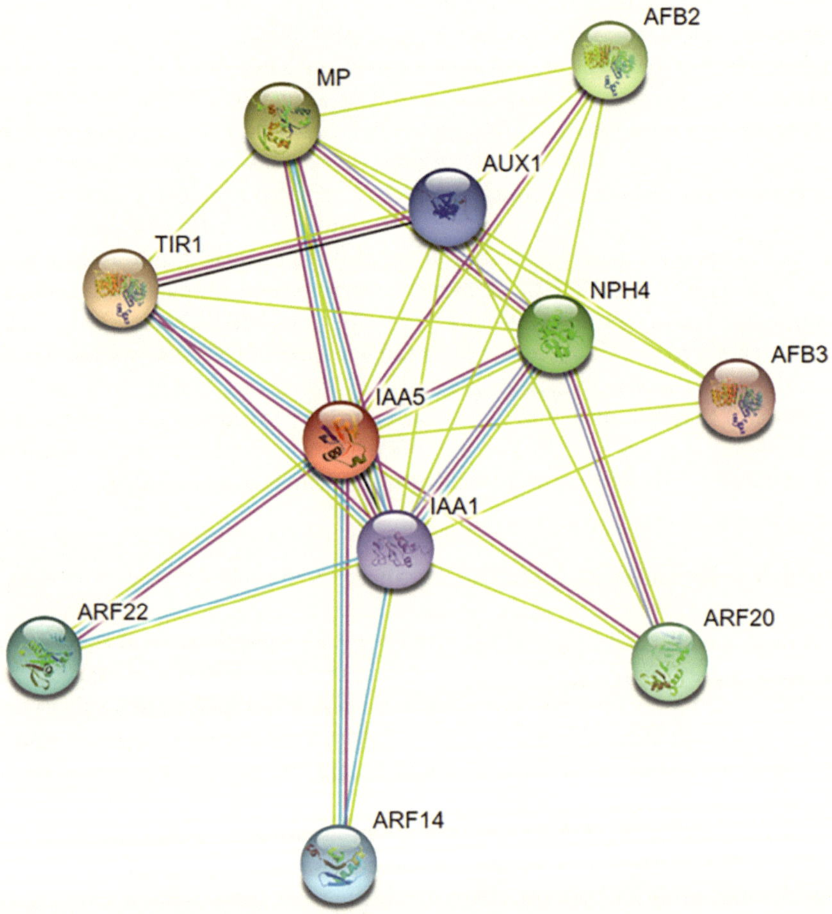

Biological Process (Gene Ontology)

GO-term	description	count in network	strength
GO:0010152	Pollen maturation	3 of 17	2.64
GO:0010311	Lateral root formation	3 of 45	2.22
GO:0009734	Auxin-activated signaling pathway	11 of 193	2.15
GO:0048527	Lateral root development	4 of 106	1.97
GO:0048443	Stamen development	3 of 90	1.92
GO:0009630	Gravitropism	2 of 72	1.84
GO:0021700	Developmental maturation	4 of 192	1.72
GO:0048364	Root development	5 of 454	1.44
GO:0009908	Flower development	4 of 415	1.38
GO:0048827	Phyllome development	4 of 430	1.37
GO:0048367	Shoot system development	5 of 723	1.24
GO:0099402	Plant organ development	6 of 881	1.23
GO:0009791	Post-embryonic development	6 of 1237	1.08

Figure 3.
The IAA5 and IAA1 interaction with other pathways in primed seedlings in Tomato.

important pathways regulated by IAA were root and lateral root development (**Figure 3**). Auxins induce root formation in seedlings [108–111]. Primed plants have an induced auxin level to form root systems. Earlier studies have shown that copper and zinc treatments led to significant alterations in the transcript levels of several auxin transporter-related genes, such as *Solyc01g007010.3*, which encodes a RING-type E3 ubiquitin ligase. Copper and Zinc treatment of seeds induces auxin levels in seedlings and helps root formation in a microgravity environment [54]. Other hormones related to root development were also upregulated by priming tomato seedlings. Cu and Zn induce a priming mechanism in the seed to make the seeds more sustainable under microgravity environmental stress during growth and development.

2.1 Engineering the future of cosmic cultivation: Advanced breeding paradigms for space agriculture

The formidable challenge of establishing sustainable agriculture in space demands revolutionary breeding strategies that transcend terrestrial limitations. Microgravity, ionizing radiation, and resource scarcity necessitate a fundamental reimagining of crop development protocols through synergistic integration of cutting-edge biotechnologies and classical breeding wisdom.

At the molecular frontier, CRISPR-Cas9 systems enable surgical precision in editing stress-responsive genetic networks. Targeting key regulators of redox homeostasis (APX, CAT), phytohormone crosstalk (ARF, DELLA), and epigenetic memory (ROS1, DME) allows creation of plants with engineered resilience [49, 50, 111–113]. This genome-editing revolution is amplified by high-throughput phenotyping platforms that screen mutant libraries under simulated space conditions.

Marker-assisted selection has evolved into ultra-precise genomic prediction models incorporating machine learning algorithms. By analyzing millions of single-nucleotide polymorphism (SNP) markers associated with radiation tolerance (RAD51, DMC1) and microgravity adaptation (LAZY1, ARG1), breeders can accelerate development of space-optimized cultivars [105–111]. The integration of multi-omics pipelines reveals previously hidden adaptation mechanisms—from stress-induced metabolite fluctuations to alternative splicing variants that enhance extraterrestrial survival.

A groundbreaking approach combines primed epigenomes with fixed genetic improvements. Chemically or environmentally primed seeds exhibit transgenerational stress memory through DNA methylation patterns (NRPD1, CMT3) and histone modifications (HDA6, SUVH4), while CRISPR-edited loci provide permanent trait enhancements [111–114]. This dual strategy creates crops capable of both immediate adaptation and long-term evolutionary success in space environments.

The next phase involves developing autonomous breeding systems utilizing AI-driven genomic selection and robotic phenotyping under artificial gravity gradients. Such closed-loop systems will be essential for establishing self-sustaining agricultural ecosystems on Mars and beyond, ensuring food security for a multiplanetary civilization.

3. Conclusion

As human exploration extends beyond Earth for prolonged periods, reliable and sustainable Advanced Life Support (ALS) systems will become essential. Plants form the cornerstone of these systems, providing food, oxygen, and waste recycling. To thrive in extraterrestrial environments, especially under microgravity conditions, it is vital to enhance plant resilience from the earliest stages of growth. Seed priming with metal ions such as copper (Cu) and zinc (Zn) has shown promising results in inducing stress memory and activating signaling pathways related to germination, root architecture, hormone balance, and defense responses. These priming strategies promote the development of gravity-resistant seedlings and can significantly improve germination rates, root establishment, and overall plant vigor in space environments. Integrating such approaches with molecular breeding techniques offers a robust path toward developing cultivars tailored for space farming, ultimately supporting human life beyond Earth.

Acknowledgements

The authors gratefully acknowledge Zerogravity Solutions for their financial support.

Author details

Kiran R. Kharat[1] and Raveendran Pottathil[2]*

1 School of Life Sciences, Mizoram University, Aizawl, Mizoram, India

2 Zero Gravity Solutions, Inc., Boca Raton, FL, USA

*Address all correspondence to: accudx@gmail.com

References

[1] Mustard JF, Murchie SL, Pelkey SM, Ehlmann BL, Milliken RE, Grant JA, et al. Hydrated silicate minerals on Mars observed by the Mars reconnaissance orbiter CRISM instrument. Nature. 2008;**454**(7202):305-309

[2] Benna M, Hurley DM, Stubbs TJ, Mahaffy PR, Elphic RC. Lunar soil hydration constrained by exospheric water liberated by meteoroid impacts. Nature Geoscience. 2019;**12**(5):333-338

[3] Nakajima S, Nagata M, Ikehata A. Mechanism for enhancing the growth of mung bean seedlings under simulated microgravity. npj Microgravity. 2021;**7**(1):26

[4] Haveman N, Paul A, Ferl R. Plant biology and a new approach to space farming. In: In-Space Manufacturing and Resources. Germany: Wiley-VCH; 2022. pp. 67-87

[5] Colla G, Battistelli A, Proietti S, Moscatello S, Rouphael Y, Cardarelli M, et al. Rocket seedling production on the international space station: Growth and nutritional properties. Microgravity Science and Technology. 2007;**19**(5-6): 118-121

[6] Dauzart AJC, Vandenbrink JP, Kiss JZ. The effects of clinorotation on the host plant, medicago truncatula, and its microbial symbionts. Frontiers in Astronomy and Space Sciences. 2016;**3**:3

[7] Vandenbrink JP, Kiss JZ. Space, the final frontier: A critical review of recent experiments performed in microgravity. Plant Science. 2016;**243**:115-119

[8] Buchheim JI, Matzel S, Rykova M, Vassilieva G, Ponomarev S, Nichiporuk I, et al. Stress related shift toward inflammaging in cosmonauts after long-duration space flight. Frontiers in Physiology. 2019;**10**:85

[9] Zeng D, Cui J, Yin Y, Xiong Y, Liu M, Guan S, et al. Metabolomics analysis in different development stages on SP0 generation of rice seeds after spaceflight. Frontiers in Plant Science. 2021;**12**:700267

[10] Barker R, Lombardino J, Rasmussen K, Gilroy S. Test of *Arabidopsis* space transcriptome: A discovery environment to explore multiple plant biology spaceflight experiments. Frontiers in Plant Science. 2020;**11**:147

[11] Micco M, Collie GW, Dale AG, Ohnmacht SA, Pazitna I, Gunaratnam M, et al. Structure-based design and evaluation of naphthalene Diimide G-Quadruplex ligands as telomere targeting agents in pancreatic cancer cells. Journal of Medicinal Chemistry [Internet]. 2013;**56**(7):2959-2974. DOI: 10.1021/jm301899y

[12] Driss-Ecole D, Legué V, Carnero-Diaz E, Perbal G. Gravisensitivity and automorphogenesis of lentil seedling roots grown on board the international Space Station. Physiologia Plantarum. 2008;**134**(1):191-201

[13] Yu F, Driss-Ecole D, Rembur J, Legué V, Perbal G. Effect of microgravity on the cell cycle in the lentil root. Physiologia Plantarum. 1999;**105**(1):171-178

[14] Li X, Liu F. Drought stress memory and drought stress tolerance in plants: Biochemical and molecular basis. In: Drought Stress Tolerance in Plants. Vol. 1. Cham: Springer International Publishing; 2016. pp. 17-44

[15] Liu X, Quan W, Bartels D. Stress memory responses and seed priming correlate with drought tolerance in plants: An overview. Planta. 2022;**255**(2):45

[16] Ramírez DA, Rolando JL, Yactayo W, Monneveux P, Mares V, Quiroz R. Improving potato drought tolerance through the induction of long-term water stress memory. Plant Science. 2015;**238**:26-32

[17] Walter J, Nagy L, Hein R, Rascher U, Beierkuhnlein C, Willner E, et al. Do plants remember drought? Hints towards a drought-memory in grasses. Environmental Experimental Botany. 2011;**71**:34-40

[18] Bruce TJA, Matthes MC, Napier JA, Pickett JA. Stressful "memories" of plants: Evidence and possible mechanisms. Plant Science. 2007;**173**(6):603-608

[19] Sinclair BJ, Roberts SP. Acclimation, shock and hardening in the cold. Journal of Thermal Biology. 2005;**30**(8):557-562

[20] Chen K, Fessehaie A, Arora R. Selection of reference genes for normalizing gene expression during seed priming and germination using qPCR in Zea mays and *Spinacia oleracea*. Plant Molecular Biology Reporter. 2012;**30**(2):478-487

[21] Savvides A, Ali S, Tester M, Fotopoulos V. Chemical priming of plants against multiple abiotic stresses: Mission possible? Trends in Plant Science. 2016;**21**(4):329-340

[22] Hilker M, Schmülling T. Stress priming, memory, and signalling in plants. Plant Cell and Environment. 2019;**42**(3):753-761

[23] Hilker M, Schwachtje J, Baier M, Balazadeh S, Bäurle I, Geiselhardt S, et al. Priming and memory of stress responses in organisms lacking a nervous system. Biological Reviews. 2016;**91**(4):1118-1133

[24] Sen A, Puthur JT. Seed priming-induced physiochemical and molecular events in plants coupled to abiotic stress tolerance: An overview. In: Priming-Mediated Stress and Cross-Stress Tolerance in Crop Plants. USA: Elsevier; 2020. pp. 303-316

[25] de Micco V, Aronne G, de Pascale S. Effect of simulated microgravity on seedling development and vascular differentiation of soy. Acta Astronautica. 2006;**58**(3):139-148

[26] de Micco V, de Pascale S, Paradiso R, Aronne G. Microgravity effects on different stages of higher plant life cycle and completion of the seed-to-seed cycle. Plant Biology. 2014;**16**:31-38

[27] Wolff SA, Coelho LH, Zabrodina M, Brinckmann E, Kittang AI. Plant mineral nutrition, gas exchange and photosynthesis in space: A review. Advances in Space Research. 2013;**51**(3):465-475

[28] Ričkienė A. Space plant biology research in Lithuania. Endeavour. 2012;**36**(3):117-124

[29] Wyatt SE, Kiss JZ. Plant tropisms: From Darwin to the international space station. American Journal of Botany. 2013;**100**(1):1-3

[30] Merkys A, Laurinavicius R. Complete cycle of individual development of *Arabidopsis thaliana* (L.) Heynh. plants on board the Salyut-7 orbital station. Doklady Akademii Nauk SSSR. 1983;**271**:509-512

[31] Dayanandan P. Gravitational biology and space life sciences: Current status and implications for the Indian space

programme. Journal of Biosciences. 2011;**36**(5):911-919

[32] Paul AL, Zupanska AK, Schultz ER, Ferl RJ. Organ-specific remodeling of the *Arabidopsis* transcriptome in response to spaceflight. BMC Plant Biology. 2013;**13**(1):112

[33] Manzano AI, Herranz R, Manzano A, van Loon JJWA, Medina FJ. Early effects of altered gravity environments on plant cell growth and cell proliferation: Characterization of morphofunctional nucleolar types in an *Arabidopsis* cell culture system. Frontiers in Astronomy and Space Sciences. 2016;**3**:2

[34] Volkmann D, Behrens HM, Sievers A. Development and gravity sensing of cress roots under microgravity. Die Naturwissenschaften. 1986;**73**(7):438-441

[35] Laurinavichius RS, Yaroshyus AV, Marchyukaytis A, Shvyaghdene DV, Mashinskiy AL. Metabolism of pea plants grown under space flight conditions. USSR Space Life Sciences Digest. 1986;**4**:23-25

[36] Kuang A. Cytochemical localization of reserves during seed development in *Arabidopsis thaliana* under spaceflight conditions. Annals of Botany. 1996;**78**(3):343-351

[37] Musgrave M. Changes in *Arabidopsis* leaf ultrastructure, chlorophyll and carbohydrate content during spaceflight depend on ventilation. Annals of Botany. 1998;**81**(4):503-512

[38] Moore R, McClelen CE, Fondren WM, Wang CL. Influence of microgravity on root-cap regeneration and the structure of columella cells in Zea mays. American Journal of Botany. 1987;**74**(2):218-223

[39] Hilaire E, Peterson BV, Guikema JA, Brown CS. Clinorotation affects morphology and ethylene production in soybean seedlings. Plant & Cell Physiology. 1996;**37**(7):929-934

[40] Klymchuk DO, Kordyum EL, Vorobyova TV, Chapman DK, Brown CS. Changes in vacuolation in the root apex cells of soybean seedlings in microgravity. Advances in Space Research. 2003;**31**(10):2283-2288

[41] Kuang A, Xiao Y, McClure G, Musgrave ME. Influence of microgravity on ultrastructure and storage reserves in seeds of *Brassica rapa* L. Annals of Botany. 2000;**85**(6):851-859

[42] Campbell WF, Salisbury FB, Bugbee B, Klassen S, Naegle E, Strickland DT, et al. Comparative floral development of mir-grown and ethylene-treated, earth-grown super dwarf wheat. Journal of Plant Physiology. 2001;**158**(8):1051-1060

[43] van de Poel B, van der Straeten D. 1-aminocyclopropane-1-carboxylic acid (ACC) in plants: More than just the precursor of ethylene! Frontiers in Plant Science. 2014;**5**:640

[44] Fenner M. Seed and seedling ecology. In: Handbook of Functional Plant Ecology. USA: CRC Press; 1999. pp. 589-648

[45] Aronne G. Seasonal dimorphism in the mediterranean *Cistus incanus* L. subsp. incanus. Annals of Botany. 2001;**87**(6):789-794

[46] Janmohammadi M, Dezfuli PM, Sharifzadeh F. Seed invigoration techniques to improve germination and early growth of inbred line of maize under salinity and drought stress. General and Applied Plant Physiology. 2008;**34**(3-4):215-226

[47] Nasir MW, Yasmeen A, Imran M, Zoltan T. Seed priming to alleviate drought stress in cotton. Journal of Environmental and Agricultural Sciences. 2019;**21**:14-22

[48] Merkys AJ, Laurinavičius RS, Švegždiene DV. Plant growth, development and embryogenesis during Salyut-7 flight. Advances in Space Research. 1984;**4**(10):55-63

[49] Sabatini S, Heidstra R, Wildwater M, Scheres B. SCARECROW is involved in positioning the stem cell niche in the *Arabidopsis* root meristem. Genes & Development. 2003;**17**(3):354-358

[50] Ferl RJ, Paul AL. The effect of spaceflight on the gravity-sensing auxin gradient of roots: GFP reporter gene microscopy on orbit. npj Microgravity. 2016;**2**(1):15023

[51] Eisvand HR, Tavakkol-Afshari R, Sharifzadeh F, Maddah Arefi H, Hesamzadeh Hejazi SM. Effects of hormonal priming and drought stress on activity and isozyme profiles of antioxidant enzymes in deteriorated seed of tall wheatgrass (*Agropyron elongatum* host). Seed Science and Technology. 2010;**38**(2):280-297

[52] Bagheri N, Alizadeh O, Zadeh SS, Aref F, Ordookhani K. Evaluation of auxin priming and plant growth promoting Rhizobacteria on yield and yield components of wheat under drought stress. Eurasian Journal of Biosciences. 2019;**13**(2):711-716

[53] Ulfat A, Majid SA, Hameed A. Hormonal seed priming improves wheat (*Triticum aestivum* L.) field performance under drought and non-stress conditions. Pakistan Journal of Botany. 2017;**49**(4):1239-1253

[54] Kharat KR, Pottathil R. Chemically defined elicitors activate priming in tomato seedlings. Plant Signaling & Behavior. 2022:e2095143

[55] Legué V, Yu F, Driss-ecole D, Perbal G. Effects of gravitropic stress on the development of the primary root of lentil seedlings grown in space. Journal of Biotechnology. 1996;**47**(2-3):129-135

[56] Hell R, Stephan UW. Iron uptake, trafficking and homeostasis in plants. Planta. 2003;**216**(4):541-551

[57] Musgrave ME, Kuang A, Xiao Y, Stout SC, Bingham GE, Briarty LG, et al. Gravity independence of seed-to-seed cycling in *Brassica rapa*. Planta. 2000;**210**(3):400-406

[58] Kuang A. Influence of microgravity on ultrastructure and storage reserves in seeds of *Brassica rapa* L. Annals of Botany. 2000;**85**(6):851-859

[59] Ma JF, Furukawa J. Recent progress in the research of external Al detoxification in higher plants: A mini review. Journal of Inorganic Biochemistry. 2003;**97**(1):46-51

[60] Sun YL, Hong SK. Effects of citric acid as an important component of the responses to saline and alkaline stress in the halophyte *Leymus chinensis* (Trin.). Plant Growth Regulation. 2011;**64**(2):129-139

[61] Balmer A, Pastor V, Glauser G, Mauch-Mani B. Tricarboxylates induce defense priming against bacteria in *Arabidopsis thaliana*. Frontiers in Plant Science. 2018;**9**:1221

[62] Kharat KR, Pottathil R. Defense priming in okra using chemically defined elicitors. International Journal of Biochemistry Research & Review. 2021;**30**(9):40-50

[63] Mittler R, Vanderauwera S, Gollery M, van Breusegem F. Reactive

oxygen gene network of plants. Trends in Plant Science. 2004;**9**(10):490-498

[64] Paul AL, Zupanska AK, Ostrow DT, Zhang Y, Sun Y, Li JL, et al. Spaceflight transcriptomes: Unique responses to a novel environment. Astrobiology. 2012;**12**(1):40-56

[65] Sugimoto M, Oono Y, Gusev O, Matsumoto T, Yazawa T, Levinskikh MA, et al. Genome-wide expression analysis of reactive oxygen species gene network in Mizuna plants grown in long-term spaceflight. BMC Plant Biology. 2014;**14**(1):4

[66] Paul AL, Popp MP, Gurley WB, Guy C, Norwood KL, Ferl RJ. *Arabidopsis* gene expression patterns are altered during spaceflight. Advances in Space Research. 2005;**36**(7):1175-1181

[67] Gadjev I, Vanderauwera S, Gechev TS, Laloi C, Minkov IN, Shulaev V, et al. Transcriptomic footprints disclose specificity of reactive oxygen species signaling in *Arabidopsis*. Plant Physiology. 2006;**141**(2):436-445

[68] Swanson S, Gilroy S. ROS in plant development. Physiologia Plantarum. 2010;**138**(4):384-392

[69] Suzuki N, Koussevitzky S, Mittler R, Miller G. ROS and redox signalling in the response of plants to abiotic stress. Plant, Cell & Environment. 2012;**35**(2):259-270

[70] Sng NJ, Kolaczkowski B, Ferl RJ, Paul AL. A member of the CONSTANS-like protein family is a putative regulator of reactive oxygen species homeostasis and spaceflight physiological adaptation. AoB Plants. 2019;**11**(1):ply075

[71] Contento AL, Kim SJ, Bassham DC. Transcriptome profiling of the response of *Arabidopsis* suspension culture cells to Suc starvation. Plant Physiology. 2004;**135**(4):2330-2347

[72] Loreti E, Poggi A, Novi G, Alpi A, Perata P. A genome-wide analysis of the effects of sucrose on gene expression in *Arabidopsis* seedlings under anoxia. Plant Physiology. 2005;**137**(3):1130-1138

[73] Bailly C. ROS in seed germination. In: Advances in Botanical Research. Netherlands: Elsevier; 2023. pp. 177-204

[74] Forti C, Ottobrino V, Bassolino L, Toppino L, Rotino GL, Pagano A, et al. Molecular dynamics of pre-germinative metabolism in primed eggplant (*Solanum melongena* L.) seeds. Horticulture Research. 2020;**7**(1):87

[75] Wang Y, Sun X, Peng J, Li F, Ali F, Wang Z. Regulation of seed germination: ROS, epigenetic, and hormonal aspects. Journal of Advanced Research. 2024;**71**:107-125

[76] Deyong Z, Jie C, Yishu Y, Meng Z, Shan S, Xin G, et al. Effects of space flight on expression of key proteins in rice leaves. Rice Science. 2020;**27**(5):423-433

[77] Chen J, Le XC, Zhu L. Metabolomics and transcriptomics reveal defense mechanism of rice (*Oryza sativa*) grains under stress of 2,2′,4,4′-tetrabromodiphenyl ether. Environment International. 2019;**133**:105154

[78] Kumari A, Parida AK. Metabolomics and network analysis reveal the potential metabolites and biological pathways involved in salinity tolerance of the halophyte *Salvadora persica*. Environmental and Experimental Botany. 2018;**148**:85-99

[79] Ma Y, Cheng Z, Wang W, Sun Y. Proteomic analysis of high yield rice

variety mutated from spaceflight. Advances in Space Research. 2007;**40**(4):535-539

[80] Zeng D, Cui J, Yin Y, Zhang M, Shan S, Yao LM, et al. Proteomic analysis in different development stages on SP0 generation of rice seeds after space flight. Life Sciences in Space Research (Amst). 2020;**26**:34-45

[81] Kruse CPS, Meyers AD, Basu P, Hutchinson S, Luesse DR, Wyatt SE. Spaceflight induces novel regulatory responses in *Arabidopsis* seedling as revealed by combined proteomic and transcriptomic analyses. BMC Plant Biology. 2020;**20**(1):237

[82] Choi W, Barker RJ, Kim S, Swanson SJ, Gilroy S. Variation in the transcriptome of different ecotypes of *Arabidopsis thaliana* reveals signatures of oxidative stress in plant responses to spaceflight. American Journal of Botany. 2019;**106**(1):123-136

[83] Chandler JO, Haas FB, Khan S, Bowden L, Ignatz M, Enfissi EMA, et al. Rocket science: The effect of spaceflight on germination physiology, ageing, and transcriptome of *Eruca sativa* seeds. Life. 2020;**10**(4):49

[84] Yurkevich OY, Samatadze TE, Levinskikh MA, Zoshchuk SA, Signalova OB, Surzhikov SA, et al. Molecular cytogenetics of *Pisum sativum* L. grown under spaceflight-related stress. BioMed Research International. 2018;**2018**:1-10

[85] Musgrave ME, Kuang A. Plant reproductive development during spaceflight. Advances in Space Biology and Medicine. 2003;**9**:1-23

[86] Hoson T. Plant growth and morphogenesis under different gravity conditions: Relevance to plant life in space. Life. 2014;**4**(2):205-216

[87] Wiszniewska A. Priming strategies for benefiting plant performance under toxic trace metal exposure. Plants. 2021;**10**(4):623

[88] Salmi ML, Roux SJ. Gene expression changes induced by space flight in single-cells of the fern *Ceratopteris richardii*. Planta. 2008;**229**(1):151-159

[89] Salmi ML, Bushart TJ, Roux SJ. Autonomous gravity perception and responses of single plant cells. Gravitational and Space Research. 2011;**25**(1):6-13

[90] Paul AL, Daugherty CJ, Bihn EA, Chapman DK, Norwood KLL, Ferl RJ. Transgene expression patterns indicate that spaceflight affects stress signal perception and transduction in *Arabidopsis*. Plant Physiology. 2001;**126**(2):613-621

[91] Stutte GW, Monje O, Hatfield RD, Paul AL, Ferl RJ, Simone CG. Microgravity effects on leaf morphology, cell structure, carbon metabolism and mRNA expression of dwarf wheat. Planta. 2006;**224**(5):1038-1049

[92] Jaenisch R, Bird A. Epigenetic regulation of gene expression: How the genome integrates intrinsic and environmental signals. Nature Genetics. 2003;**33**(S3):245-254

[93] Li E, Beard C, Jaenisch R. Role for DNA methylation in genomic imprinting. Nature. 1993;**366**(6453):362-365

[94] Mohandas T, Sparkes RS, Shapiro LJ. Reactivation of an inactive human X chromosome: Evidence for X inactivation by DNA methylation. Science 1979. 1981;**211**(4480):393-396

[95] Finnegan EJ, Peacock WJ, Dennis ES. DNA methylation, a key regulator of plant development and other processes.

Current Opinion in Genetics & Development. 2000;**10**(2):217-223

[96] Litt MD, Hornstra IK, Yang TP. In vivo footprinting and high-resolution methylation analysis of the mouse hypoxanthine phosphoribosyltransferase gene 5′ region on the active and inactive X chromosomes. Molecular and Cellular Biology. 1996;**16**(11):6190-6199

[97] Stroud H, Greenberg MVC, Feng S, Bernatavichute YV, Jacobsen SE. Comprehensive analysis of silencing mutants reveals complex regulation of the *Arabidopsis* methylome. Cell. 2013;**152**(1-2):352-364

[98] Zilberman D, Gehring M, Tran RK, Ballinger T, Henikoff S. Genome-wide analysis of *Arabidopsis thaliana* DNA methylation uncovers an interdependence between methylation and transcription. Nature Genetics. 2007;**39**(1):61-69

[99] Sudan J, Raina M, Singh R. Plant epigenetic mechanisms: Role in abiotic stress and their generational heritability. 3 Biotech. 2018;**8**(3):172

[100] Mauch-Mani B, Mauch F. The role of abscisic acid in plant–pathogen interactions. Current Opinion in Plant Biology. 2005;**8**(4):1037, 409-414

[101] Williams BP, Gehring M. Stable transgenerational epigenetic inheritance requires a DNA methylation-sensing circuit. Nature Communications. 2017;**8**(1):2124

[102] Link BM, Busse JS, Stankovic B. Seed-to-seed-to-seed growth and development of *Arabidopsis* in microgravity. Astrobiology. 2014;**14**(10):866-875

[103] Zhou M, Sng NJ, LeFrois CE, Paul AL, Ferl RJ. Epigenomics in an extraterrestrial environment: Organ-specific alteration of DNA methylation and gene expression elicited by spaceflight in *Arabidopsis thaliana*. BMC Genomics. 2019;**20**(1):205

[104] Califar B, Sng NJ, Zupanska A, Paul AL, Ferl RJ. Root skewing-associated genes impact the spaceflight response of *Arabidopsis thaliana*. Frontiers in Plant Science. 2020;**11**:239

[105] Han X, Kahmann R. Manipulation of phytohormone pathways by effectors of filamentous plant pathogens. Frontiers in Plant Science. 2019;**10**:822

[106] Mähönen AP, Ten TK, Siligato R, Smetana O, Díaz-Triviño S, Salojärvi J, et al. PLETHORA gradient formation mechanism separates auxin responses. Nature. 2014;**515**(7525):125-129

[107] Sato EM, Hijazi H, Bennett MJ, Vissenberg K, Swarup R. New insights into root gravitropic signalling. Journal of Experimental Botany. 2015;**66**(8):2155-2165

[108] Blancaflor EB, Fasano JM, Gilroy S. Mapping the functional roles of cap cells in the response of *Arabidopsis* primary roots to gravity 1. Plant Physiology. 1998;**116**(1):213-222

[109] Leyser O. Auxin signaling. Plant Physiology. 2018;**176**(1):465-479

[110] Gray WM, Kepinski S, Rouse D, Leyser O, Estelle M. Auxin regulates SCFTIR1-dependent degradation of AUX/IAA proteins. Nature. 2001;**414**(6861):271-276

[111] Salehin M, Bagchi R, Estelle M. SCF TIR1/AFB-based auxin perception: Mechanism and role in plant growth and development. The Plant Cell. 2015;**27**(1):9-19

[112] Erdoğan İ, Cevher-Keskin B, Bilir Ö, Hong Y, Tör M. Recent developments in CRISPR/Cas9 genome-editing technology related to plant disease resistance and abiotic stress tolerance. Biology (Basel). 2023;**12**(7):1037

[113] Rai GK, Khanday DM, Kumar P, Magotra I, Choudhary SM, Kosser R, et al. Enhancing crop resilience to drought stress through CRISPR-Cas9 genome editing. Plants. 2023;**12**(12):2306

[114] Chen F, Chen L, Yan Z, Xu J, Feng L, He N, et al. Recent advances of CRISPR-based genome editing for enhancing staple crops. Frontiers in Plant Science. 2024;**15**:1478398